T0188619

Honeypots

A New Paradigm to
Information Security

Honeypots
A New Paradigm to Information Security

R.C. Joshi

*Department of Electronics and
Computer Engineering*

IIT, Roorkee, India

Anjali Sardana

*Department of Electronics and
Computer Engineering*

IIT, Roorkee, India

CRC Press
Taylor & Francis Group
an **informa** business
www.crcpress.com

6000 Broken Sound Parkway, NW
Suite 300, Boca Raton, FL 33487
270 Madison Avenue
New York, NY 10016
2 Park Square, Milton Park
Abingdon, Oxon OX 14 4RN, UK

Science Publishers
Enfield, New Hampshire

Published by Science Publishers, P.O. Box 699, Enfield, NH 03748, USA
An imprint of Edenbridge Ltd., British Channel Islands

Email: _info@scipub.net_   Website: _www.scipub.net_

Marketed and distributed by:

CRC Press	6000 Broken Sound Parkway, NW Suite 300, Boca Raton, FL 33487
Taylor & Francis Group	270 Madison Avenue New York, NY 10016
an **informa** business	2 Park Square, Milton Park
www.crcpress.com	Abingdon, Oxon OX 14 4RN, UK

Copyright reserved © 2011

ISBN 978-1-57808-708-2

CIP data will be provided on request.

Printed in the United States of America

Preface

Security has received much focus and attention since the past decade. Any threat to the network security can lead to heavy financial, economic and other loses. So, protection of network against security threats has become a crucial prerequisite for any organization. Since architecture of the Internet was designed with an aim to provide maximum functionality, it has some inherent weaknesses and incapability which result in successful origin and execution of attacks. Internet failures can be accidental or intentional. The design of Internet has been made to an extent of controlling accidental failures. But intentional attack by malicious users has no answer in original Internet design. Honeypot makes use of the vulnerabilities to trap such intentional attackers and study attacks.

Honeypot is an important security technology used to understand and combat attacks. It is designed with deliberate vulnerabilities, which is exposed to a public network. The goal is first to lure intruders away from the real system, and secondly to closely monitor the intruder to study the exploits which are used.

OBJECTIVES

This book is written to give the reader a well-rounded understanding of honeypot in wired and wireless networks. Going through the book, the reader will see that it addresses honeypots from many angles. Moreover, the book is exhaustive, systematic and user friendly. It has a simple methodological approach and case studies to enhance the practical understanding of the subject along with strong theoretical foundation. The purpose of this book is to provide theoretical ground and practical survey of both principles and practice of Honeypots. The book deals with the practical and latest technology in Information Security, Honeypots, including honeytokens, honeynets and honeyfarms. It further discusses role of honeypots in various attack scenarios like denial of service, virus, worms, phishing, etc. and elaborates on virtual honeypots and forensics. Practical Implementations as well as current state of research has been discussed.

The wide scope of knowledge that this book brings will help one become acquainted with many aspects of honeypots. This book also has career advancement in mind by covering all the objectives a honeypot can serve in the market. The subject, and therefore this book, draws on a variety of disciplines. Attempt has been made to make the book self contained.

INTENDED AUDIENCE

The wide scope of the book benefits anyone who has to administer, secure, hack, practise, understand or research on security technologies. The book is intended for both an academic and professional audience. As a text book, it is intended as a one-semester course in Honeypots at undergraduate levels/post graduate levels in Computer Science, Information Technology, Network Security, Information Science and Management. The book serves as basic reference volume for researchers and is suitable for self study. It should also be useful to practitioners and consultants in security industry.

REVIEW QUESTIONS, ANALYTICAL EXERCISES AND CASE STUDIES

The book includes a number of short questions for review at the end of each chapter. They can be used to check whether the reader actually understood the materials of the section.

There are several long exercises at the end of each chapter that ask the reader to prove results left unproved in the text or supplementary results that are useful for their own sake. The readers are logically expected to study, solve and reason them out towards the end of each chapter.

PLAN OF THE BOOK

The book is organized into a number of chapters. The chapter starts with basic concepts for novice readers, moving to intricacies of the subject and ends with current state of research and results.

It begins with key concepts and terminology, which sets the stage for rest of the section. Some of the classic and popular commercially available honeypots have been explained and their installation had been given. The motive is to make reader of the book familiar to the technology and its interface. It then discusses the role of honeypots under different phases of an attack for attacks like spam, fishing, worms and virus and DDoS attacks. The concepts of static, virtual, and dynamic honeypots has been elaborated with case studies. It also describes virtual honeypot VMware and its significance has been focused through case study. Wireless honeypots, a very recent concept has been discussed in details. It is followed by applications of Honeypot in real—life scenarios including Tactical Battlefield, Information Warfare, Forensic Analysis, Network Surveillance etc. The section ends with Anti-Honeypot technology which demonstrates that every Honeypot solution has a fingerprint that is ultimately detected by the attacker and honeypot is compromised in due course of time. To overcome this anti honeypot technology, honeypot should upgrade itself before compromised be anti honeypots. Thus there is a warfare going on between attackers and honeypots which will never end.

PREREQUISITES

The reader of this book is assumed to have a working knowledge of the basics of networking. An understanding of the following OSI layers is taken for granted: Transport layer, Network layer, Data Link layer. Any relevant concepts besides these layers will be introduced as the discussion progresses. The analysis requires prior knowledge of probability theory and statistics.

Contents

1

Honeypots

This chapter begins by outlining the trends in network attacks over the past three decades that have motivated the use of honeypots. It further defines the essence of honeypots and honeytokens and the reason to use them. It details out the ways in which honeypots can be categorized and their security value in the network. The chapter further elaborates the various locations where honeypots can be deployed in network and the ways in which each placement affects the overall security. Finally, the risks and trade-offs involved in operating honeypots are presented.

The advancement of internet technology has made the world look like one interconnected network of millions of digital information generating systems. Amenities like ATMs, NetBanking, Debit & Credit cards, even e-mails, chats, etc., have "eased" everyone's life. These amenities have increased the risk of a breach of information manifold. With fast progress in networking, storage and processing technologies and easy availability of information, the threats over exposure of digital data have also increased.

The objective of security includes protection of network, information and property from theft, corruption, or alteration, while allowing the information and property to remain accessible and productive to its intended users. Many security solutions have been proposed in order to protect the unauthorized access of information and internet resources. Various effective mechanisms are in place to detect the attacks during the event and for its analysis after the attack has happened. But normally, one is unable to analyze systems properly after they are compromised and it becomes difficult to collect detailed information about the hacker and his motives. Obtaining information regarding an attack apriori, before an actual violation, has always been challenging.

Honeypot is a technology that works to gather such apiori knowledge about attacks by luring the hackers to attack it. It acts as bait that entices suspected users to attempt to commit an anticipated malicious act. The suspected attacker's movements are then monitored and analyzed.

"You can catch more flies with honey rather than vinegar"

The term honeypot was first coined during the Cold War as a spying technique. Year 1990 marks the beginning of use of honeypot concept in the field of information security with the publication of "The cuckoos-Egg" and "An Evening with Berferd in which a Cracker is Lured, Endured and Studied". The greatest challenge for an organization is to know who their enemies are, how they might attack, when they might attack, what the enemies do once they compromise a system, and, perhaps most important, why they attack. Honeypots are able to improvise this information with the organization to a great extent.

1.1 BACKGROUND

Gone are the days when one could be content with merely an antivirus running on his desktop. In this so called "Internet Friendly" world one lives in now, breach of valuable information is not a tough nut to "crack" since weaknesses exist in system architecture, system configuration, application design, implementation configuration, and operations.

Internet Security has become very important in this era, the following frauds will explain for themselves:

> *Case 1*
> For several months, beginning in the fall of 1996, two credit union employees worked together to alter credit reports in exchange for financial payment [2]. As part of their normal responsibilities, the employees were permitted to alter credit reports based on updated information the company received. However, the employees intentionally misused their authorized access to remove negative credit indicators and add fictitious indicators of positive credit to specific credit histories in exchange for money. The total amount of fraud loss from their activities exceeded $215,000. The risk exposure to the credit union was incalculable.

Case 2

UBS PaineWebber
In March 2002, a "logic bomb" deleted 10 billion files in the computer systems of the international financial services company, UBS PaineWebber. Logic bomb is a malicious code implanted on a target system and configured to execute after a designated period of time or on the occurrence of a specified system action. This incident affected over 1300 of the company's servers throughout the United States. The company sustained losses of approximately $3 million, the amount required to repair damage and reconstruct the deleted files. Investigations by law enforcement professionals and computer forensic professionals revealed the logic bomb had been planted by a disgruntled employee who had recently quit the company because of a dispute over the amount of his annual bonus.

Case 3

Monster job site hacked
In August 2007, the online job site Monster.com suffered a security breach that reportedly resulted in the theft of the confidential information from some 1.3 million job seekers. That figure was later revised to "millions".

Hackers stole information from the US online recruitment site's password-protected CV library by using credentials taken from Monster clients. They launched the attack using two servers at a web-hosting company in the Ukraine, combined with a botnet. The compromised computers had been infected with a malicious software program known as Infostealer.Monstres.

The company first learned of the problem on 17 August, when investigators with internet security company Symantec told Monster that it was under attack.

Attacks existed during the 1980s and early 1990s, but at that time they were not viewed publicly as being high-profile security incidents. They became costly and crucial only when the Internet became a mainstream medium due to the phenomenal growth in late 1990's. Trojans, viruses, worms, DDoSs have all developed over time and are evolving at a much faster rate than the antivirus-signatures [5]. Lack of appropriate security has lead to a huge number of disasters as shown in Figure 1.1 and the graph is rising. According to a study conducted by the Computer Security Institute in 2004: These attacks caused a loss of more than $26,064,050 million US dollars.

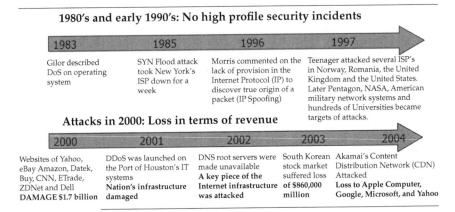

1980's and early 1990's: No high profile security incidents

1983	1985	1996	1997
Gilor described DoS on operating system	SYN Flood attack took New York's ISP down for a week	Morris commented on the lack of provision in the Internet Protocol (IP) to discover true origin of a packet (IP Spoofing)	Teenager attacked several ISP's in Norway, Romania, the United Kingdom and the United States. Later Pentagon, NASA, American military network systems and hundreds of Universities became targets of attacks.

Attacks in 2000: Loss in terms of revenue

2000	2001	2002	2003	2004
Websites of Yahoo, eBay Amazon, Datek, Buy, CNN, ETrade, ZDNet and Dell DAMAGE $1.7 billion	DDoS was launched on the Port of Houston's IT systems **Nation's infrastructure damaged**	DNS root servers were made unavailable **A key piece of the Internet infrastructure was attacked**	South Korean stock market suffered loss of $860,000 million	Akamai's Content Distribution Network (CDN) Attacked **Loss to Apple Computer, Google, Microsoft, and Yahoo**

Figure 1.1 Network Attack Incidents.

Since 2004, incidents have deliberately not been widely publicized, as they have shifted to the sensitive field of economic crime, and harm the victim's reputation. Cyber-extortion has become common, in which the victims are threatened to be put offline until they pay, with typical targets being credit card processing companies [6]. Since the Internet downtime is very damaging in terms of finances, and the reputation of online companies is very important, most victims choose to pay. In January 2006, the www.milliondollarhomepage.com, a British teenager's novel advertising idea to earn $1 million in four months, became famous very quickly and drew instant attention of cyber-extortionists, who bombarded the website with intense DoS [7–10]. They initially asked $5,000 to avert them and finally $50,000 to stop them. The text-to-speech translation application running in the Sun Microsystem's Grid Computing system was disabled on the very day it was launched with a DoS attack in March, 2006. The very latest DDoS incident that occurred on 29 May, 2009 turned down Internet in various parts of China, putting few million Chinese Internet users into trouble. Thus, the timeline shows that over the time, attacks have become more frequent, sophisticated and effective in obstructing the availability of services and causing significant damages.

In maximum cases of such cyber frauds, one is unaware about who his enemy is. Due to this restriction, one is unable to protect himself from hackers.

An intruder to a system can be referred to as a hacker or a cracker.

A hacker is basically someone who hacks a system because he finds it interesting. Hackers can be categorized as either whitehat (also called crackers) or blackhat. Both of them know how to penetrate into the system. But the motives of penetrating are different for both of them.

Whitehat's aim is to know a system's loopholes so as to make the system more secure against attacks.

Blackhats' make use of this knowledge for personal gain and other selfish and unethical purposes. Blackhats may begin with a port scan, then may deface a website or launch a Denial of Service attack or even leave a Backdoor for future access to the system [4, 11]. The only way to learn about the Blackhat community than to watch them in action is to record step-by-step as they attack and compromise a system and also to know about their motives after they compromise a system.

A cracker hacks the system because he wants to access the organization's system. In any case, hackers and crackers are both intruders and can be classified as external or internal, i.e., outsiders or insiders.

External Intruders: are outsiders for an organization's network. They attack web servers, email servers and may also attempt to go bypass the firewall to attack machines on the internal network. Outside intruders may come from the Internet, dial-up lines, physical break-ins, or from a partner (vendor, customer, reseller, etc.) network that is linked to an organization's corporate network as in case 3.

Internal Intruders: They belong to the organization's network. They are legitimate users for a network. These include users who misuse privileges or who attempt to get higher rights or use other user's privileges (refer to case 1). Internal intruders are often overlooked. Most of the security breaches (almost 80%) are done by insiders.

According to the SANS Institute, there are seven management errors that lead to computer security vulnerabilities [12]:

1. Pretending the problem will go away.
2. Authorizing reactive, short-term fixes. So problems re-emerge rapidly.

3. Failing to realize how much money their information and organizational reputations are worth.
4. Relying primarily on a firewall and IDS.
5. Failing to deal with the operational aspects of security: make a few fixes and then not allow the follow through necessary to ensure the problems stay fixed.
6. Failing to understand the relationship of information security to the business problem—they understand physical security but do not see the consequences of poor information security.
7. Assigning untrained people to maintain security and not providing the training or the time to make it possible to do the job.

Traditionally, tools applied for security of the information have been purely defensive. These include access control tests, firewalls, intrusion detection systems (IDS), encryption techniques, etc., The strategy followed the classical security paradigm of "Protect, Detect and React": try to protect the network as best as possible, detect any failures in the defense, and then react to those failures. The problem with this approach is that the attacker has the initiative, being always one step ahead. Hence it is hard to effectively secure a communication network. The continuous race between the hackers and the security-experts to outdo each other is simply ceaseless. Since the attacks are being modified daily, current security technologies cannot face them.

Thus the technology is now focused to change this defensive procedures (that are point solution or that provide only perimeter security) and try to learn more about hackers, their attack patterns and behavior. The concept of electronic decoy (honeypots), i.e., information system resources, which are used to be probed, attacked and exploited, are quite a lot helpful in this field.

1.1.1 History and Evolution of Honeypots

In 1991, a number of publications expounded on concepts that were to be foundations of today's honeypot development. Two publications in particular stood out: These were Clifford Stoll's "The Cuckoos Egg" and Bill Cheswick's "An Evening with Berferd". In 1997, Fred Cohen released the Deception Toolkit. The Deception Toolkit is one of the original and landmark Honeypots. It is generally a collection of PERL scripts designed for UNIX systems that emulate a variety of known vulnerabilities. In fact, 1998 the following year gave us the first commercial Honeypot implementation called CyberCop sting3. CyberCop Sting is a component of the CyberCop intrusion protection software family which runs on NT. Another commercial Honeypot implementation that came out in the same year was NetFacade4. As with CyberCop Sting, it creates a simulated network of hosts, with simulated IP addresses, running seemingly vulnerable services but in a much larger scale. 1998 was a particularly busy year for Honeypot development because after CyberCop and NetFacade, the windows Honeypot "Back Officer Friendly" came out. In 1999 a group of people led by Lance Spitzner [13] decided to form the Honeynet Project. The honeynet project is a non-profit group dedicated to researching the blackhat community and to share their work to others. In 2002, another related initiative was born out of the Honeynet Project this time involving the whole security community to further Honeypot related research [13, 14]. This initiative is what one calls now the Honeynet Research Alliance [4]. The Research Alliance is composed of groups around the world interested in Honeypot research. This setup provided a means for Honeypots to be deployed around the world and provided a venue for sharing tools and techniques used in Honeypot research. In 2003, several important Honeypot tools were introduced by these organizations such as Snort-Inline [5], Sebek [15], and advanced virtual honeynets [16]. Since then newer and newer projects are coming on honeypots for both wired and wireless environments.

1.2 HONEYPOTS

A *honeypot* is

> "A program that takes the appearance of an attractive service, set of services, an entire operating system or even an entire network, but is in reality a tightly sealed compartment built to lure and contain an attacker."

Like a hidden surveillance camera, a honeypot monitors and logs every action an attacker makes including access attempts, keystrokes, files accessed and modified, and processes executed.

They can do everything from detecting encrypted attacks in IPv6 networks to capturing the latest in on-line credit card fraud.

Lance Spitzner, the founder of the Honeynet Project, defines a honeypot as:

> "A honeypot is an information system resource whose value lies in unauthorized or illicit use of that resource" [17].

A honeypot runs emulated operating system and services which act as a trap luring a hacker to compromise the system while in reality, a honeypot logs all the means the hacker uses to compromise the system and also the actions of the hacker after he compromises the system. A more sophisticated option for networked organizations is to use a Honeynet which consists of a number of honeypots distributed over a network of Production Systems.

1.2.1 Generic Honeypot Model

A honeypot is set up on a network for the sole purpose of being attacked. It is designed with deliberate vulnerabilities, which is exposed to a public network. No production value is assigned to a honeypot. Honeypots, thus, are not supposed to receive any legitimate traffic. Therefore, any traffic destined to a honeypot is most probably an ongoing attack and can be analyzed to reveal vulnerabilities targeted by attackers.

A honeypot comprises:

1. *Honeypot production system:* It is not a true production system, but a prey for intruders. This provides the honey—files and fake system resources for the intruder to play with. Automatic responses to intruder's actions are set up to show the honeypot as a true production system.
2. *Firewall:* Firewalls provide logs about how an intruder is attempting to get into a Honeypot [18]. Firewall is set up to log all packets going to the Honeypot system, as there should be no legitimate reason for traffic going to or from the Honeypot.
3. *Monitoring unit:* It is a threat evaluation unit that monitors network and/or system activities for malicious activities or policy violations and produces reports to a Management Station. Reviewing the order, sequence, time stamps and type

of packets used by an intruder to gain access to Honeypot and the keystrokes, system accesses, files changed, etc., help to identify the tools, methodology being used by the intruders and their intentions (vandalism, data theft, remote launch point search etc.). An IDS can do the work of a monitoring unit.

4. *Alert Unit*: Honeypot should be able to generate alerts by email or pager to send notification about traffic going to or from the honeypot to the administrator to let him review intruder activity WHILE it's happening.

5. *Logging Unit:* This unit provides efficient storage for all the firewall and system logs and of the traffic going between the firewall and the honeypot system.

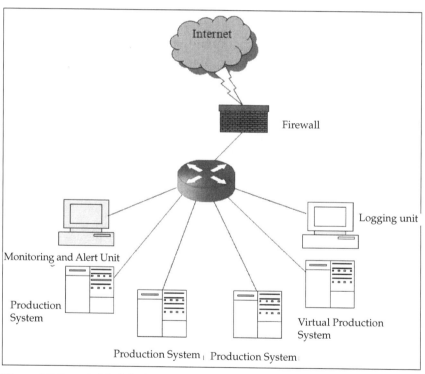

Figure 1.2 Generic Model of Honeypot.

An Illustration can be:

The default installation of WinNT 4 with IIS 4 can be hacked using several different techniques and play the role of honeypot in the network. A standard intrusion detection system can then be used to log hacks directed against the machine, and further track what the intruder attempts to do with the system once it is compromised. Install special software designed for this purpose. It has the advantage of making it look like the intruder is successful without really allowing them access. Also, on WinNT, it is possible to rename the default "administrator" account, then create a dummy account called "administrator"with no password. WinNT allows extensive logging of a person's activities, so this honeypot will track users attempting to gain administrator access and exploit that access.

1.3 HONEYPOTS VS. FIREWALLS AND INTRUSION DETECTION SYSTEMS

Although honeypots are used along with Firewall and IDS but they all should not be confused to be the same [19]. The very essence of each of these is different. Intrusion detection systems or firewalls provide only primitive security.

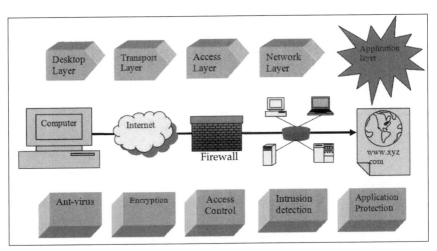

Figure 1.3 Placement of Firewall and IDS in network.

1.3.1 Firewalls

The use of firewalls at the border of the network can control the flow of traffic between the local network and the Internet [18, 20]. Based on the characteristics of the network traffic, to include requested services, source and destination addresses, and individual users, a firewall will make a decision on whether to allow the traffic to pass through the network. Firewalls can also be utilized on individual host based systems.

However, there are recognized shortfalls with the use of firewalls to protect a network.

The shortcomings associated with a firewall include the following:

1. The firewall cannot protect against attacks that bypass it.
2. The firewall at the network interface does not protect against internal threats.
3. The firewall cannot protect against the transfer of virus laid in files and programs.
4. In certain cases high volumes of network traffic may overwhelm the network monitoring capability of the firewall resulting in the possible passing of malicious traffic between networks.

1.3.2 Intrusion Detection Systems

An Intrusion Detection System (IDS) detects and alerts on possible malicious events within a network. IDS sensors may be placed at various points throughout the network. An IDS is normally signature based, i.e., it will look for predefined signatures of bad events. These signatures or rules normally reside in a database associated with the IDS. As soon as the IDS detect attack signatures, it updates the firewall's filtering rules. The use of IDS and firewalls provide a level of security protection to the system administrator.

The use of IDS as a network security device also has several shortcomings:

1. A network based IDS must be able to see all network traffic of the network that it is protecting. If a network uses a switch (most do nowadays) a sniffer will not be able to see all the network traffic. This usually means that one would deploy a network based IDS at the gateway only, i.e., on his Internet connection. However, this does not protect him from internal attacks.

2. Modern networks are so fast that an Intrusion Detection system has a hard time keeping up.

3. IDS suffer from Data Overload. They tend to generate an extremely large volume of alerts. This volume makes it time consuming, resource intensive, and costly to analyze and review all the alerts the NIDS generate. For example, IDS deployed in an organization can generate over 100,000 alerts a day. As many of these alerts are false alerts or false positives, administrators begin to ignore the technology and the alerts.

4. It can be difficult for some NIDS technologies to discover or identify unknown attacks or behaviour. This leaves the organization vulnerable to new attacks.

5. IDS require resource intensive hardware to keep up with organization's activity and traffic. The faster the network and the more data one has, the bigger the IDS will have to be to keep up.

6. More and more organizations are moving to encryption, all of the data is encrypted. This is due to security issues, regulation, and encryption technologies are more widely available (SSH, SSL, IPSec). However, these same technologies blind the NIDS so they can no longer monitor the network traffic.

1.3.3 Honeypots

Honeypots take care of these tradeoffs and provide Small Data Sets.

1. They only collect data when someone or something is interacting with them. Organizations that may log thousands of alerts a day may only log a hundred alerts with honeypots. This makes the data honeypots collect much easier to manage and analyze.

2. Honeypots dramatically reduce false positives. Any activity with honeypots is by definition unauthorized, making it extremely effective at detecting attacks.

3. They can easily identify and capture new attacks against them. Any activity with the honeypot is an anomaly, making new or unseen attacks easily stand out.

4. Honeypots require minimal resources, even on the largest of networks. A simple Pentium computer can monitor literally millions of IP addresses.

5. It does not matter if an attack is encrypted, the honeypot will capture the activity.

The strategy behind honeypots is to shut the intruders safely from production systems and to obtain information about the intruders by logging their actions [21]. Figure 1.4 shows an example of how honeypots can be deployed in a network for protection purposes. In this example, Honeypot-A simulates a system without any firewall or Intrusion Detection System (IDS); Honeypot-B simulates a vulnerable service like Ftp or Telnet service to catch attention of Blackhat; Honeypot-C and Honeypot-D simulates other systems in an organization network in order to lure Blackhat away from the real systems.

Most honeypots are installed with firewalls. The difference in the firewall and a honeypot is that honeypot works in the reverse direction. It allows all traffic to come in but blocks all outgoing traffic. Most honeypots are installed inside network firewalls and is a means of monitoring and tracking hackers.

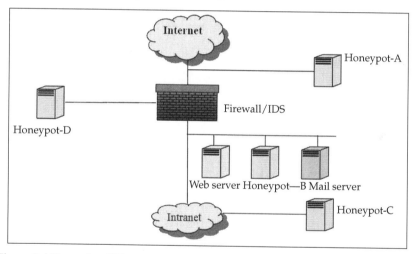

Figure 1.4 Example of Honeypot deployment in network.

Main difference between honeypot based detection and IDS lies in that a honeypot detects compromises by virtue of system activities, while an IDS compares intrusion mode with known a signature. So honeypot are more effective while detecting new or unknown attacks.

1.4 CLASSIFICATION OF HONEYPOTS

There are several types of honeypots, which can be grouped into four major categories.

1.4.1 Based on Usage

Production Honeypots

These are used to protect organizations. The purpose of a production honeypot is to help mitigate risk in an organization. Production honeypots can significantly reduce the risk of intrusion by uncovering vulnerabilities and alert administrators of attacks. They add value to security measures of an organization. They are not able to keep the Blackhat hackers out. The organization still need to depend on their security policies, procedures and best practices such as disabling unused services, patch management, implementing security mechanisms such as firewall, intrusion detection systems, anti-virus and secure authentication mechanisms to keep the Blackhat community out of the organization's IT infrastructure. They can be compared with 'law enforcement' body dealing with the Blackhat community. Commercial organizations use production honeypots to protect their network.

Research Honeypots

These are used to research the threats organizations face, and how to better protect against those threats. These types of honeypots are more focused on researching the actions of the intruder by using a number of different configurations to lure them in. These honeypots do not add direct value to a specific organization. Instead they are used to research the threats organizations face, and how to better guard against those threats. They can be compared with 'counter-intelligence' body. Their job is to gain information about Blackhat community. This information is then used to protect against those threats. Traditionally, commercial organizations do not use research honeypots. Instead, organizations such as universities, government, military, or security research organizations use them. Also, research honeypots are excellent tools for capturing automated attacks, such as auto-rooters or worms. Since these attacks target entire network blocks, research honeypots can quickly capture these attacks for analysis. The Honeynet Project, for example is a volunteer, non-profit security research organization that uses honeypots to collect information on cyber threats.

When to use which type of Honeypot?

In order to design a honeypot system, one needs to define the goals for wanting one in the first place. There are many questions to be answered before one starts, including the following:

1. What is the primary reason for wanting the honeypot?
2. What OS environment is required to be emulated with the honeypot?
3. What servers or services are required to be emulated?
4. Does one want to monitor internal threats, external threats, or both?
5. Does one want to offer unpatched systems as bait, or is concerned with only successful exploits against fully patched systems?

The answers to these questions essentially define whether one shall need a production or research honeypot, and the way to configure it.

1.4.2 Based on Level of Interaction

Low Interaction Honeypots

They are characterized by its minimal interaction with the hacker and emulate fake services [22]. These types of honeypots typically emulate a specific service like ftp or http. There are no real operating systems or services running on them, they are just emulations running above operating system layer. They are much simpler to deploy and maintain, but log only a limited amount of information regarding the hacker's activities. Since they are running above OS layer, it saves the system from the attacker control. The maximum amount of damage the attacker can do is that he can take down that honeypot emulation. Low-interaction honeypots are helpful in identifying attackers IP addresses. Some examples of commercial low-interaction honeypots are Honeyd and Specter.

Medium Interaction Honeypots

They try to combine advantages of both low interaction honeypots and high interaction honeypots. They are more advanced than low interaction honeypots but less advanced than high interaction honeypots. They will give a certain feedback when they are queried upon. These honeypots neither have real operating system environment nor do they implement all details of application protocol. They have a layer of virtualization. They just provide responses that are awaited by the hacker. These responses are used to lure hackers so that they send their payload. On receiving the payload, security personnel extract the shell code from it for investigation and forensic analysis. After detailed analysis, these honeypots are then used to emulate the actions that shell code was supposed to perform. After performing these actions to download Malware from serving location, it is stored locally or submitted somewhere else for analysis point of view. These honeypots are quite complex, time consuming in their deployment and require huge effort and thorough knowledge of protocols, application services and security to build them. They are also vulnerable to

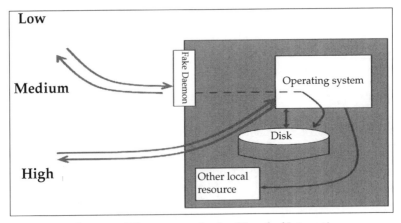

Figure 1.5 Classification of Honeypot on basis of Level of Interaction.

higher risks. Medium interaction honeypots can be regarded as homemade honeypots customized to accommodate certain needs of organizations or individuals.

High Interaction Honeypots

They give a real operating system to attack upon, thus luring the hacker, while on the other hand exposing the system to ample risk. Since high interaction honeypots provide complete access, the possibility to accumulate information about the attack as well as the attractiveness of the honeypot increases a lot, so they are specially used for research purposes. High-interaction honeypots are very difficult to deploy because multiple tools are used to make them run. Once a high-interaction honeypot is successfully deployed and managed, it can be very helpful in discovery of new exploits, worms, viruses and vulnerabilities. Examples are ManTrap and Honeynets.

When to use which type of Honeypot?

Interaction measures the amount of activity attacker can have with a honeypot. The order from top to bottom indicates increasing effectiveness while also increasing complexity, risk and difficulty in deploying. The same is expressed in the following table:

Table 1.1 Low Interaction vs. High Interaction.

Factors	Low interaction	Medium interaction	High interaction
Degree of involvement	Low	Medium	High
Real operating system	No	No	Yes
Installation	Easy	Difficult	More difficult
Maintenance	Easy	Easy	Time consuming
Risk	Low	Medium	High
Compromised wished	No	No	Yes
Need control	No	NO	Yes
Knowledge to run	Low	Low	High
Knowledge to develop	Low	High	Mid-High
Data gathering	Limited	Medium	Extensive
Interaction	Emulated services	Requests	Full control

Low Interaction Honeypots are used when: There is insufficient hardware to set up a honeynet and the risk of another type of honeypot is not acceptable.

The purpose is:

- identify scans and automated attacks
- fool script kiddies
- distract hackers from important systems
- collect attack signatures and trends

Medium Interaction Honeypots are used when: The small amount of risk involved is acceptable.

The purpose is:

- fool hackers to a much greater extent than done by Low Interaction Honeypots
- distract the bad guys from accessing important files

High Interaction Honeypots are used when: The purpose is to observe the intruders activities and behavior, observe a real compromise, and for IRC conversations.

Need material for research and training in:

- Artifact analysis
- Forensic analysis

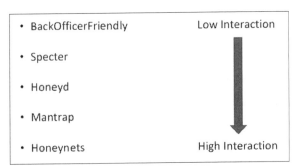

Figure 1.6 Some Commercially Available of Honeypots.

1.4.3 Based on Hardware Deployment Type

Physical Honeypots

It is a single machine running a real OS and real services, where honeypot is connected to a network and is accessible through a single IP address. Physical honeypots are always connected with the concept of high interaction honeypots. Physical honeypots are less pratical in real scenarios due to limited view of their single IP address and high cost involved in maintaining a farm of physical honeypots. Honeynets are example of physical honeypots.

Virtual Honeypots

They are usually implemented using a single physical machine that host several virtual honeypots. Virtual honeypots are more cost effective in monitoring large IP address spaces and emulating large IP addresses at the same time. An example of virtual honeypot is AGROS.

1.4.4 Based on Role of Honeypot

Server Side

Conventional honeypots are passive by design and do not intiante any traffic unless compromised. Server side honeypots are useful in detecting new exploits, collecting malware, and enriching research of threat analysis. Examples are low-interaction honeypots, honeyd etc.

Client Side

These are active honeypots for client side attacks. Client side attacks represent attacks that target vulnerable client applications, such as web servers, when client interacts with malicious servers. The aim of these honeypots is to search and detect these malicious servers. An example of a client side honeypot is Strider Honey Monkey.

   The classification of honeypots can be summarized by Figure 1.7.

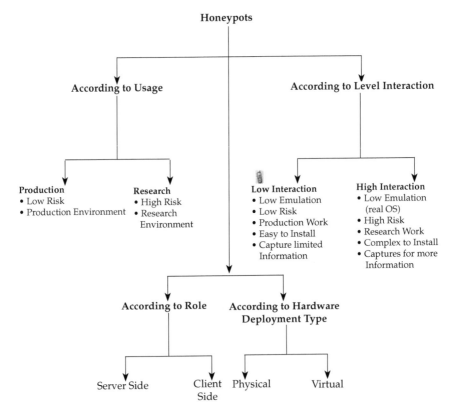

Figure 1.7 Classification of Honeypots.

Table 1.2 Classification of Honeypots.

Interaction Level		Honeypot Types		
Low	**High**	**Real OS**	**VM**	**Emulated**
Mimics only IP stack	Can mimic all layers of OSI model	Has all layers of OSI model	Has all layers of OSI model	Mimics OSI model layers depending on interaction level
Easier to set up	Harder to set up	Medium to hard to set up	Harder to set up	Can be easy or hard to set up
Lowest time commitment	High time commitment	High time commitment	High time commitment	Low to high time commitment
Redeployment after compromise is easy	Redeployment after compromise can be harder	Redeployment after compromise can be hard	Redeployment after compromise is very easy	Redeployment after compromise is easiest

1.5 PLACEMENT OF HONEYPOTS

Honeypots can be placed externally as well as internally. Conceptually they can be placed at three main locations in an organization.

- External, i.e., facing the Internet
- Internal, i.e., behind the firewall
- On the DMZ

Honeypot placement is usually discussed in its relationship to the perimeter-protecting firewall. Each location has its advantages and disadvantages, depending on the honeypot goals. Figure 1.8: Placement of honeypots.

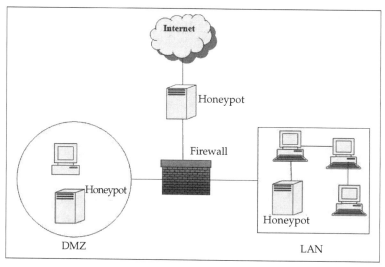

Figure 1.8 Placement of Honeypot.

1.5.1 External Placement

The honeypots are placed outside the network perimeter when one wants to capture the most malicious hackers. Honeypots connected directly to the Internet can be freely compromised and probed, as shown in Figure 1.9. This is the easiest setup for single personal, home-based and research honeypots. With external placement, there is no firewall in front of the honeypot. The honeypot and production network share the same public IP address subnet. It

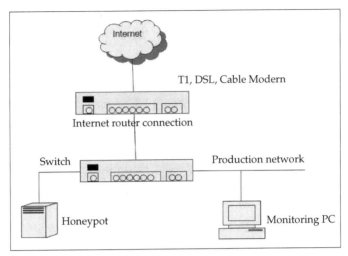

Figure 1.9 External Placement of Honeypot.

requires one or more public IP addresses. If there is only one public IP address and hub is available the public IP address is given to the honeypot and the monitoring station is set up without an IP address.

In promiscuous mode with a non-Windows IP stack, the management workstation should be able to capture traffic headed to and from the honeypot. If switch is available the monitoring workstation can be set up in port mirroring mode. The use of a switch offers some protection over a hub, but without a firewall or some other sort of defense, this type of Honeynet represents the largest risk to the production network.

The lack of a secondary router, firewall, or some other inline device makes data control very difficult. Many people start out with this method because it is the easiest to set up. They have good intentions: they plan to plug in the honeypot only while they are actively monitoring it, or they create some sort of alert system that will page them when the honeypot is compromised. In either case, they intend to physically watch the hacker's activity, and then pull the honeypot off the network if the hacker begins to attack other targets. This sounds great, but the devil is in the details.

If honeypot is plugged only while it is being actively watched, its exposure is limited. One may see some worm attacks in progress, but being right at the honeypot during an interesting manual hack attack session may be missed.

If one is using the alert method, one may be at a location globally far off when the honeypot is compromised. In the time it takes one to travel from his current location to the honeypot, the hacker could have used it to compromise other hosts. Data control is not very risky, but lack of it is the highest legal risk to a honeypot administrator.

Placing honeypots outside the firewall reduces the risk to the internal network, but it limits their ability to emulate the production systems and generate logs, which are relevant to the internal network.

1.5.2 Internal Placement

Another common honeypot system location is inside the network, with the firewall between it and the outside world, as shown in Figure 1.10.

This placement is the best way to create an early-warning system to generate alert if any external exploits have made it past other network defenses and catch internal threats at the same time. For example, during the Blaster worm attacks, any companies that had their firewalls configured to block port 135 were essentially safe from the worm. But the worm was able to sneak past the firewall on trusted VPN links and infected mobile laptop computers. Once past the firewall, the worm was able to infect unpatched internal machines. A honeypot would at least be an early-warning system that the worm had made it past the firewall.

On the downside, if an internal honeypot is compromised, data control within the local network is difficult. A hacker or worm could use the exploited honeypot to look for additional internal hosts to compromise. One can minimize that threat by placing yet another firewall (or other inline mechanism) on the honeypot/Honeynet to limit outgoing activity (known as Reverse Firewall), or use a low-interaction honeypot.

Because the honeypot system is placed behind the firewall, administrators will need to decide what Internet traffic is directed

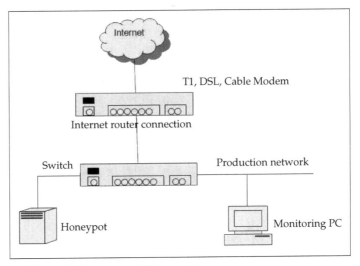

Figure 1.10 Internal Placement of Honeypot.

to the honeypot versus production assets. Will they allow any port traffic to the honeypot, or just redirect specific ports? For example, if the production network does not have a web server, the honeypot administrator might redirect any incoming HTTP requests to the honeypot instead. The use of the tarpit (a honeypot) slows down the worm's progress and benefits the local network and the Internet. It is important to decide which ports are allowed past the firewall and where the traffic should head.

In this scenario, utilizing a port mirroring switch will decrease the chance that a hacker would detect the monitoring efforts. Note that the logging computer does not need a second network interface card, because the Honeynet and the production network are one and the same.

1.5.3 DMZ Placement

Placing a honeypot on the firewall DMZ (De-militarized zone), as shown in Figure 1.11, is often the best choice for a company. It can be placed alongside other legitimate DMZ servers and provide early warning of threats located there. A router is placed between the firewall's DMZ as an added layer for data control. The Honeynet and production DMZ servers share the same logical subnet and IP address scheme. The DMZ can have public or private IP addresses.

The IDS/packet capturing computer uses the switch's port mirroring abilities in order to remain hidden. The placement of the Honeynet within the DMZ is an ideal location for many entities, but it is also the most complex placement model. Additionally, because it is located on the DMZ, it is not the best early-warning indicator for an internal network compromise.

Placing honeypots inside DMZ is a better idea as not all traffic is allowed to pass through the firewall. But then it also requires that all other system in DMZ are hardened and secured up. By placing them inside the DMZ, they can easily emulate the servers that are freely accessible to the public domains. This also increases the security of the production environment because of the limited access to internal network from the DMZ. But if the organization wants to detect hackers entering their private network, they may like to place the honeypot in their intranet environment.

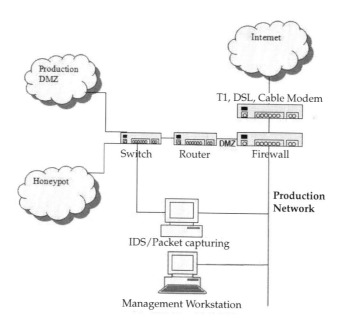

Figure 1.11 DMZ Placement of Honeypot.

1.5.4 On the Whole

Although being inside the network honeypot introduces some risks, especially when the internal network is not secured against the honeypot through additional security mechanisms, they can emulate the production systems as well as can monitor the attacks made from inside the network. They give proper logs of all the activities and can be easily integrated with other security technologies to get the best output. The honeypot in front of the firewall will generate lots of traffic such as port scans, exploits and unauthorized attempts. For honeypots that are placed before the firewall, it will pose an increased risk to the organization as attackers would attempt to attack and interact with this honeypot. By placing the honeypots before the firewall, one can experience both advantage and disadvantage of honeypots. Disadvantage would be that, it will increase the publicity of organization to hackers, thus bringing in more attackers to compromise the honeypot. Advantage would be that when attackers, especially script-kiddies, who do not know how to spoof their originating IP address realize that they are attacking a honeypot, they might be frightened off. But, this honeypot would also receive a lot of false positives from the wild and it would definitely overwhelm the administrators, who are taking care of this honeypot.

1.6 VARIETIES OF HONEYPOTS: HONEYTOKENS, HONEYPAGES, HONEYNETS AND HONEYFARMS

1.6.1 Honeytokens

They are a digital bait of any form. These are honeypots that are not computer systems. A honeytoken is any object without legitimate production value placed only as an early warning mechanism. Honeytokens can be placed on honeypots or on regular production servers. For example, a honeytoken can be an inactive hoax user account called Administrator, without any permission (It is common for the actual Administrator account on Windows computers to be renamed to some other nondescript login name as a way to impede hackers and automated hacking tools). If anyone tries to log in to the computer or network using the hoax Administrator user account, an alert is generated. Another example of honeytokens is fake username/passwords in the user database. These users do not exist in the real world, and hence are not expected to be logging in to the application. If the application sees these credentials being used, it immediately recognizes that the user database has been compromised.

1.6.2 Honeypages

They are obscure web pages sprinkled in the web site. They have no legitimate purpose. Normal users would never reach these pages. While normal users would never see this, an attacker who analyzes the source code, or a vulnerability scanner that spiders the site would see these and follow the link. When the page is accessed, it points to the intruder.

1.6.3 Honeynets

Two or more honeypots on a network form a honeynet. Typically, a honeynet is used for monitoring a larger and/or more diverse network in which one honeypot may not be sufficient. The concept of honeynet was proposed by Lance Spitzner in 1999.

1.6.4 Honeyfarms

Instead of deploying large numbers of honeypots, or honeypots on every network, honeypots are deployed at a consolidated location. This single network of honeypots becomes the honeypot farm, a dedicated security resource [23]. Attackers are then redirected to the farm, regardless of what network they are on or probing. It is basically, a centralized collection of honeypots and analysis tools.

1.7 VALUE OF HONEYPOTS

Honeypots are often the best tool for a security defense.

1.7.1 Low False-Positives

Security logs with many false-positives are considered to contain a lot of noise. False-positives are very common in intrusion detection systems (IDSs) and firewalls, as are false-negatives, to a lesser extent. Much effort is spent trying to decrease noise coming from firewalls and IDSs [24]. Often, the noise is so high that administrators give up reading and analyzing their logs, decreasing the value of the security device.

In comparison, honeypots have no legitimate production value and should never be accessed by anyone but the honeypot administrator. Any honeypot traffic, outside the expected administrative traffic, is probably malicious. Any traffic leaving the honeypot is malicious.

Note: If a honeypot is used internally, it is not uncommon for the honeypot to detect and report non-malicious broadcast traffic. Broadcast traffic can be in the form of Address Resolution Protocol (ARP) packets and Windows NetBIOS broadcasts. Honeypot network segments should be designed to filter normal broadcasts away from the honeypot.

The low noise ratio in the captured data is considered to have high value. What a honeypot captures should always be investigated. If one wants to get involved with computer and network security only when a real compromise attempt is taking place, then one should monitor honeypots.

1.7.2 Early Detection

The low occurrence of false-positives and false-negatives naturally leads to the rapid detection of legitimate threats. Some administrators use a honeytoken to ensure early detection. Regardless of how the honeypot detects an active exploit, it can immediately generate alert to the attempted compromise. One can respond quickly, close the security hole, and minimize the damage. The same principle applies on a larger scale.

SQL Slammer worm [1]

During the early morning hours of January 25, 2003, honeypots were among the first security solutions to detect the SQL Slammer worm. Slammer attacked vulnerable Microsoft SQL Server 2000 software with a buffer overflow on UDP port 1434 and infected more than 200,000 computers in its first ten minutes of release. It brought down a major banking ATM network and caused Denial of Service (DoS) attacks across large portions of the Internet [3, 4]. Early detection allowed an internationally coordinated effort to block port 1434 traffic headed across major backbones, stopping the worm from spreading even farther. By the time most of us awoke, the biggest part of the threat was over. The worm was being quickly eradicated, detection signatures were available, and exploited networks were in cleanup mode. This was due in large part to the early detection work by honeypots and IDSs.

1.7.3 New Threat Detection

Because every connection on a honeypot is a legitimate threat, previously unknown attacks are found just as quickly as known attack vectors. For example, at least two major zero-day exploits were first discovered and documented by honeypots. New hacking methods, while not necessarily zero-day, are discovered routinely by honeypots. There are even research tools, like Honeycomb [21, 25], that allow brand-new threats discovered by a honeypot to automatically generate an IDS signature. Unlike a virus scanner or signature-detecting IDS, honeypots are excellent at detecting new threats.

Honeypots can capture everything associated with the hacker, including all network packets, uploaded malware, chat communications, and typed commands. This allows the administrator to learn what the hackers are doing and how they are doing it.

1.7.4 Defense in Depth

The defense-in-depth security paradigm states that the more defensive the tools that protecting a network are, the more successful the overall defense will be. A common use for a honeypot is to place it inside the network perimeter. If something sneaks past the firewall and IDS and ends up inside the network, there is a chance the honeypot will pick it up. A layered defense will be more likely to catch something that another solution missed. Many of today's computer viruses and worms spread by attempting to infect weakly password-protected NetBIOS shares. Scans made to ports 137 through 139 (the NetBIOS ports) [26] on the honeypot could indicate that a virus or worm has made it inside the perimeter.

It was recently discovered that hackers are setting up Internet Protocol version 6 (IPv6) stacks on machines they have exploited. The hackers then tunnel IPv6traffic inside the IPv4 traffic, creating a simple but effective virtual private network (VPN). Many IDSs and firewalls, not being designed for IPv6, can't decode the tunneled traffic and are not able to peer inside the malicious packets. A compromised honeypot in an AT&T Mexican honeynet captured hackers using IPv6 to tunnel malicious IRC traffic. The discovery of this led to an increased awareness of the importance of firewalls and IDSs in decoding IPv6 traffic. Honeypots are instrumental in knowing what the enemy is up to.

1.7.5 Other Advantages of Honeypots

1. Simplicity of the honeypot: There are no fancy algorithms to develop, state tables to maintain, or signatures to update. The simpler a technology is, the less likely is the chance of mistakes or misconfigurations.
2. They do not include any actual production services, so no one in the organization will need to access the honeypot. For this reason, any connection to the honeypot is likely a scan or attack.
3. Unlike intrusion detection systems, which only identify when and how an attacker got into the network, a honeypot is a distraction as well.
4. The data collected from honeypots is smaller but nonetheless, of high importance. They provide a lot of useful information on attacker's movement within the system. Instead of logging a GB of data a day, they can log only one MB of data a day. Instead of generating 10,000 alerts a day, they can generate only 10 alerts a day. Honeypots only capture activities of Blackhat

community, any interaction with a honeypot is most likely unauthorized or malicious activity. As such, honeypots reduce 'noise' by collecting only small data sets, but information of high importance, as it is only the Blackhat community. This means that it is much easier (and cheaper) to analyze the data a honeypot collects and to derive value from it.

5. A honeypot can also be used as an early warning system, alerting administrators of any hostile intent before the real system is compromised.

6. Honeypots can even capture tools and tactics which have never been seen before.

7. A challenge most security mechanisms face is resource limitations, or even resource exhaustion. For example, a firewall may fail because its connections table is full, it has run out of resources, or it can no longer monitor connections. This forces the firewall to block all connections instead of just blocking unauthorized activity. An Intrusion Detection System may have too much network activity to monitor, perhaps hundreds of megabytes of data per second. When this happens, the IDS sensor's buffers become full, and it begins dropping packets. Its resources have been exhausted, and it can no longer effectively monitor network activity, potentially missing attacks. Honeypots require minimal resources, they only capture bad activity. This means an old Pentium computer with 128MB of RAM can easily handle an entire class B network sitting off an OC-12 network. Thus honeypots do not have problems of resource exhaustion.

8. Unlike other security technologies (such as firewall and IDS), honeypots do not require the latest cutting-edge technology, vast amounts of RAM or chip speed, or large disk drives.

9. Unlike most security technologies (such as IDS systems) honeypots work fine in encrypted or IPv6 environments. It does not matter what the bad guys throw at a honeypot, the honeypot will detect and capture it.

10. Honeypots can collect in-depth information that few, if any other technologies can match.

1.8 RISKS AND TRADEOFFS

Different types of honeypots have different levels of risk. Few of them are:

1. Honeypots are worthless if no one attacks them.
2. Honeypots have a narrow field of view. They only see what activity is directed against them. If an attacker breaks into ones network and attacks a variety of systems, the honeypot will be blissfully unaware of the activity unless it is attacked directly. If the hacker has identified the honeypot for what it is, he can now avoid that system and infiltrate organization, with the honeypot never knowing he got in.
3. Honeypots can be used as a launching platform to attack other machines, meaning that, if it is successfully broken into, the attacker can use this to begin other attacks.
4. Honeypots encourage an aggressive atmosphere and add risk to a network.
5. If the honeypot is attacked, the administrator should prevent major changes to the honeypot, because any noticeable changes will make the attacker more suspicious.
6. Honeypots suffer from fingerprinting. Fingerprinting is when a hacker can identify the true identity of a honeypot because it has certain expected characteristics or behaviors. For example, a honeypot may emulate a Web server. Whenever a hacker connects to this specific type of honeypot, the Web server responds by sending a common error message using standard HTML. This is the exact response one would expect for any Web server. However, the honeypot has a mistake in it and misspells one of the HTML commands, such as spelling the word length as legnht. This misspelling now becomes a fingerprint for the honeypot, since any hacker can quickly identify it because of this error in the Web server emulation. An incorrectly implemented honeypot can also identify itself. If a blackhat identifies an organization using a honeypot on its internal networks, he could spoof the identity of other production systems and attack the honeypot. The honeypot would detect these spoofed attacks, and falsely alert administrators that a production system was attacking it, sending the organization

on a wild goose chase. Meanwhile, in the midst of all the confusion, an attacker could focus on real attacks.

7. Fingerprinting is an even greater risk for research honeypots. A system designed to gain intelligence can be devastated if detected. An attacker can feed bad information to a research honeypot as opposed to avoiding detection. This bad information would then lead the security community to make incorrect conclusions about the blackhat community.

8. Honeypots add complexity. In security, complexity is bad: it leads to increased exposure to exploits.

Due to these disadvantages, honeypots do not replace existing security mechanisms. However, they add value by working with existing security mechanisms.

1.9 KEY ISSUES AND CHALLENGES

Honeypots have proven to be extremely successful to provide zero "false positive" results. Three main purposes for which honeypots are used today are detection, prevention, and information gathering. In contrast to classical security methods which collect data after the system is attacked by Blackhat community, honeypots offer a more systematic approach to study attack patterns and general vulnerability assessment. It also helps us learn about the tools, tactics and motive of the Blackhat community, which help us to develop new strategies to secure networks and to defend against new kinds of attacks.

Potential applications of honeypots include commercial enterprises, government agencies, surveillance and communication in battlefield and in production and research. Honeypots are useful tools, however there are some limitations associated with their uses. One problem is the issue of liability. Intentionally placing an insecure machine on a network and allowing it to get compromised can put an organization into a very awkward position, if the attacker then uses that system to launch attacks on other organizations' or individuals' systems. Therefore, limitations have to be placed on honeypots' out-bound network connections in order to prevent abuse at the hands of attackers. Honeypots are not the solution for computer crimes [27]. Honeypots are hard to maintain and they need operators with deep knowledge about computer and network security. Remember, honeypots are not a solution. Instead, they are a tool. Their value depends on what the goal is, from early warning and detection to research.

Another key challenge is to maintain its opaqueness as high as possible. Honeypot should not reveal its presence during process of interaction with the suspected user. During interaction with the honeypot, a suspected user is proposed to be given synthetic data instead of actual sensitive data. The suspected data, if not properly constructed, may cause suspicion in the user and hence loss of opaqueness of the system. The synthetic traffic must not create suspicion in the user.

1.10 SUMMARY

In the field of security, honeypot have emerged as successful tools and technology to be used in conjunction with IDS and firewalls. The flexibility of honeypots allows them to be used for adding value to the security of an organization in production network or just for research purposes. Research honeypots contribute little to the direct security of an organization. They can be deployed outside the network, in DMZ or inside the network according to the goals and requirements and serve different purposes. Honeypots are an early detection mechanism, however, they add risk of attack in the network they are deployed.

EXERCISES

Short Answer type Questions:

1. What precautions in your view should be taken while deploying a honeypot in a network?
2. Under what circumstances honeypot should be placed in the production network?
3. Under which case would you install a honeypot, without updating certain patches?
4. Can honeypots monitor unused IP space?

5. What are the legal issues with honeypot?
6. Analytical questions.
7. Which honeypot among production or research, would you recommend for the following cases?
8. You want to imitate existing applications, services, and servers. Production assets are fully patched. The key requirement is to bait malicious user.
9. You want to protect your working assets.
10. You want to record everything a hacker does.
11. Hacker's attack is directed toward your organization and its data.
12. Organizations, such as universities, governments, or extremely large corporations are interested in learning more about threats.
13. Suggest a level of interaction of honeypot for each of the above scenarios.
14. Outline the various places at which you would place the honeypots in the network for scenarios given in 4.

Long Answer type Questions:

1. Explain the concept of honeypot giving an example from your daily life. What are the components needed to make a honeypot? Find analogies between these components and real life examples.
2. Why Honeypots are needed when firewalls and IDS are already there?
3. What would be the benefits of using honeypots in conjunction with firewalls?
4. What would be the benefits of using honeypots in conjunction with other intrusion detection system?
5. Explore the IDS "SNORT" given at www.snort.org. Outline the additional features your honeypot should have to harden the security of the network.
6. Would you suggest using firewall with honeypot? Give reason for your answer.

7. What are the main issues with honeypots in general? What are the risks associated with the internal placement of honeypots? What about placing them in DMZ?

8. Describe the relative advantages and disadvantages of placing honeypots behind firewall, outside firewall and on DMZ of a network.

9. Search for any current major project on Honeypot and explain its deployment strategies and security threats found by it.

10. What are honeytokens, honeypages, honeyfarms and honeynet?

11. What are the various ways of classifying honeypots?

12. Compare low, medium and high interaction honeypots with respect to risk, amount of data gathering, and maintenance cost.

13. What are client-side honeypots and how they are different from conventional honeypots?

14. What are major issues and challenges faced by honeypot technology?

REFERENCES

[1] K. Poulsen, "Slammer worm crashed Ohio nuke plant network," *Security Focus*, vol. 19, 2003.

[2] "CERT® Advisory CA-1996-21 TCP SYN Flooding and IP Spoofing Attacks," in http://www.cert.org/advisories/CA-1996-21.html, Sept, 1996.

[3] "CERT/CC Denial of Service," http://www.cert.org/tech_tips/denial_of_service.html, 2001.

[4] "CSI/FBI Computer Crime and Security Survey," http://i.cmpnet.com/gocsi/db_area/pdfs/fbi/FBI2004.pdf, 2004.

[5] F. Lau, S.H. Rubin, M.H. Smith, and L. Trajkovic, "Distributed denial of service attacks," in *Proceedings of IEEE International Conference on Systems, Man, and Cybernetics*, BC, Canada, 2000, pp. 2275–2280.

[6] H.F. Lipson and I. Carnegie-Mellon Univ Pittsburgh Pa Software Engineering, "Tracking and tracing cyber-attacks: Technical challenges and global policy issues", 2002.

[7] "Hackers cripple al-Jazeera sites," http://news.bbc.co.uk/2/hi/technology/2893993.stm, 2003.

[8] "South Korean markets hit by net worm," http://news.bbc.co.uk/2/hi/business/2698385.stm, 2003.

[9] "Akamai Provides Insight into Internet Denial of Service Attack", http://www.akamai.com/html/about/press/releases/2004/press_061604.html, 2004.

[10] K. Poulsen, "FBI busts alleged DDoS mafia," *Security Focus,* vol. 26, 2004.

[11] L. Garber, "Denial-of-service attacks rip the Internet", *Computer,* vol. 33, pp. 12–17, Apr 2000.

[12] C. Sima, "Security at the Next Level," *SPI Dynamics,* 2004.

[13] "The Honeypot Projects", http://old.honeynet.org/scans/scan28/.

[14] "CERT Statistics (Historical)", http://www.cert.org/stats/cert_stats.html, 2009.

[15] V.D. Gligor, "A note on the denial-of-service problem," in *Proceedings of the 1983 IEEE Symposium on Security and Privacy,* Oakland, CA, 1983, p. 139.

[16] R.T. Morris, "A weakness in the 4.2 BSD Unix TCP/IP software", *Computing science technical report,* vol. 117, 1985.

[17] L. Spitzner, "Honeytokens: The other honeypot," *Security Focus,* vol. 21, pp. 1-5, 2003.

[18] R. Oppliger, "Internet security: firewalls and beyond," *Communications of the ACM,* vol. 40, pp. 92–102, 1997.

[19] "ISS INC. :Real Secure Intrusion Detection System," http://www.iss.net.

[20] W.R. Cheswick, S.M. Bellovin, and A.D. Rubin, *Firewalls and Internet security: repelling the wily hacker*: Addison-Wesley Longman Publishing Co., Inc. Boston, MA, USA, 2003.

[21] H.D. Steps, "A Honeypot Deployment Plan," *Honeypots for Windows: Apress,* vol. Part One, pp. 35–59, 2006.

[22] "'Mafiaboy' hacker jailed," http://news.bbc.co.uk/2/hi/science/nature/1541252.stm, 2001.

[23] L. Spitzner, "Honeypot Farms," http://www.symantec.com/connect/articles/honeypot-farms, 2003.

[24] P.E. Proctor, *Practical intrusion detection handbook*: Prentice Hall PTR Upper Saddle River, NJ, USA, 2000.

[25] C. Kreibich, "Honeycomb automated ids signature creation using honeypots," http://www.cl.cam.ac.uk/cpk25/honeycomb.html, 2003.

[26] K. McCarthy, "Wanna know how BT.com was hacked?," URL: http://www.theregister.co.uk/2000/07/25/wanna_know_how_bt_com/.

[27] "Blackmailers target $1m website," http://news.bbc.co.uk/2/hi/technology/4621158.stm, 2006.

2

Commercially Available Honeypots

This chapter introduces you the simplest of the commercially available low interaction honeypot, i.e., BackOfficer Friendly. Next, a more realistic production honeypot, Specter, is discussed. Mantrap is a flexible high interaction honeypot that can be used to gather a lot of information about the attackers at the cost of risk and complexity. Finally, we provide an insight into Honeyd, an open source honeypot that can be deployed in the network and tailored according to the requirements of an organization.

In the previous chapter we have studied that honeypots are quiet useful for our organization. But the question that now arises is, which honeypot should we use for our organization. The choice of honeypot depends on organization's requirement and the financial resources available with them. Various honeypots are available in market for commercial as well as domestic use. Each of them has its own different pros and cons. None of them is best suited for our organization. We have to choose the one that is of maximum advantage to our organization and has minimum side effects on it. In this chapter we will focus on these aspects of commercially available honeypots. Although there are numerous available honeypots in market, but we will discuss about only four honeypots here viz. BackOfficer Friendly, Specter, Mantrap, Honeyd. These honeypots are the one that are most widely used in commercial world and have least adverse side effects. All the four honeypots discussed here are different from each other in one or other respect. The difference between them may be based upon their uses, their level of interaction, etc.

2.1 BACKOFFICER FRIENDLY

BackOfficer Friendly, often known as BOF is one of the simplest honeypot to use, understand and configure. It is due this simplicity of BackOfficer Friendly that it one of the excellent available tools. BackOfficer Friendly is classified as a low interaction honeypot. This honeypot runs on almost all Windows based platform, even the older ones like Windows 95, Windows 98, etc. The main advantage of BackOfficer Friendly is that it's free. One can download it free of cost from NFR Security Inc.

BackOfficer Friendly was originally designed as a response to a specific threat. It was first developed in 1998 by Marcus Ranum and the guys at Network Flight Recorder in response to the Cult of the Dead Cow's [1] , known as cDc, release of Back Orifice . It was created to detect the Back Orifice scan attempts against your computer. Back Orifice is a "backdoor" tool developed by the hacking group *Cult of the Dead Cow* and released in August 1998. Systems are infected in the normal Trojan Horse manner: a person downloads or is sent an executable from the Internet. Once the executable runs, it invisibly runs on the system, providing full access to outside hackers. Hackers regularly scan the Internet looking for people who have been compromised by this program [1]. Although BackOfficer Friendly was originally created to detect Back Orifice scan attempts only but now it has evolved to detect and alert to attacks attempted to other services also, such as Telnet, FTP, SMTP, POP3 and IMAP2, etc. Now there are seven preconfigured services on which BackOfficer Friendly can detect attacks. Whenever some guy from blackhat community attempts a connection with any of these seven services, BackOfficer Friendly listens to it on that port, generates an alert and emulates the transaction. It logs attackers IP address and operations he tries to perform. None of the services emulate a specific application or version, only the functionality of the service. Although BackOfficer Friendly has emulation capability, but it is extremely limited. Due to this limited emulation capability;

BackOfficer Friendly is classified as low interaction honeypot. It doesn't forge particular packet fields, but simply completes connection, then tears it down. It may sometime attempt to generate some meaningful replies also. Also, since additional services cannot be added or modified, BackOfficer Friendly is limited to detecting attacks only on the seven ports that BackOfficer Friendly monitors. If an attack is made against any other port, BOF is blissfully unaware of any malicious activity.

One BOF beta-tester is a police officer in the high-tech crime unit of a state police department. He says that besides the three cases he's investigated involving Back Orifice, the police production network itself is typically hit with at least one Back Orifice penetration attempt per day, reported by BOE. On the day he was contacted by Data Communications, three attempts had been made against his network.

BOF is slated to ship in April at $10 per desktop, with source code available for $25 per desktop. Network Flight Recorder costs $5500. For those not familiar with Network Flight Recorder, its founder, CEO (and a good friend of the Editor-in-Chief) is Marcus Ranum who created DEC SEAL, TIS Gauntlet and Firewall Toolkit shareware. Preliminary testing shows it to be good 'stuff', where 'stuff' is a technical term."

Case 1

The following article appeared in the Computers and Security, Vol. 16, No. 3 by Dr. Bill Hancock/Security Views before the release of BOF.

"Attacks Network Flight Recorder Inc. (www. nfr.com) will in April, 1999, release a software package that it says can be used to guard against Back Orifice, a hacker program that lets remote attackers take control of Windows machines. The package is said to intercept and block Back Orifice commands and send bogus replies to attackers. Back Officer Friendly (BOF) is agent software that runs on Windows or Unix machines. It's designed as an add-on to NFR's Network Flight Recorder intrusion detection program. BOF emulates servers for various popular services like Web, telnet, and FTP (file transfer protocol). When scans are attempted with Back Orifice (or Satan or other popular tools), BOF can send back replies. For instance, it responds to intruder E-mail requests with the message, "Bad Hacker. No Donut!"

To install BackOfficer Friendly on your computer you are required to run *nfrbofl.exe*, which available free of cost from NFR Security Inc. When prompted to start listening now, click **yes**. You can stop and restart BackOfficer Friendly later, if needed. When prompted to start BackOfficer Friendly at startup, click **yes**. You can change this option through the BackOfficer Friendly menus, if needed. After BOF installation you would like to run it, but would be wondering why.

Case 2

Attacks over a four-day period, from November 10 to November 13, 2001 were detected by BOF. The vast majority of attacks were probes looking for Microsoft IIS Web server vulnerabilities. These most likely were scripted attacks being launched from different attackers. However, the same tool was most likely being used, since consistently the same attack signatures were seen. Several probes made against FTP. These were most likely probes attempting to find servers running vulnerable versions of FTP.

BackOfficer Friendly is visible in Windows startup folder. BackOfficer Friendly does not use the Windows startup folder. Although you can change this setting from within the BackOfficer Friendly menus. Normally BackOfficer Friendly is available in system tray. To access the BackOfficer Friendly Window, you'll have to double click on the BackOfficer Friendly icon in the system tray. On double clicking a window shown below pops up.

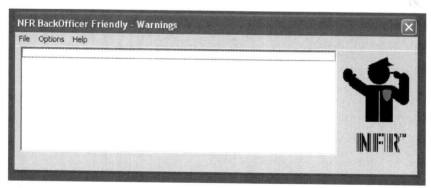

Figure 2.1 Menu for BackOfficer Friendly.

Now you can configure your BackOfficer Friendly, as you require. With the help of Option menu on the interface, as shown in figure below, you can configure it as you wish.

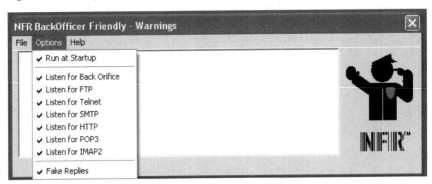

Figure 2.2 All options for configuring BOF.

If you want your honeypot to start automatically, you can select Run at Startup option else you can deselect it. You can also select all or any combination of services that you want to listen on. As you can see there is also a Fake Replies option. In default mode (Fake Replies option deselected), when an attacker communicates with any of these ports, BOF first establishes a TCP connection and then closes it with RST packet. It doesn't emulate any of the services. Thus an attacker thinks that there is no service provided. But on the contrary if the Fake Replies option has been selected, BOF enables the emulation capabilities. In this case it detects the connection and then even responds to it. Thus we can obtain more information.

After completion of installation and configuration, you can even confirm that BOF is actually monitoring the ports that you had selected. To confirm this, in the command prompt, go to the place where you have installed BackOfficer Friendly honeypot and execute the command *netstat –na*. It would show us all the ports that have an open socket listening for a connection.

Figure 2.3 using netstat –na to confirm that BOF is monitoring the ports.

As shown in figure1, when we start a BOF, we get an empty BOF window. Whenever someone attacks your computer, it is logged into this window. Since BOF has not configured to differentiate between legitimate n illegal IP address, so even if it sees any activity from same IP address it logs it. Thus we can check how the BOF works by trying to communicate with it. To do this we have to try to establish a connection with BOF using command prompt. When we open a command prompt window, we get a following window.

Figure 2.4 Command prompt.

Now first we'll see how BOF works with deselecting Fake Replies.

First of all we'll see how telnet works. We will try to open a Telnet connection with BOF by writing Telnet _._._._. In the blank spaces we have to write the IP address of computer with which connection has to be established. In this case we'll write the IP address of our own computer as shown in figure below.

Figure 2.5 Trying to establish a telnet connection.

Once we send a telnet request, a telnet window as shown below is opened.

Figure 2.6 Telnet window.

Simultaneously it generates an alert stating that a given I.P. address is trying to establish a telnet connection. An example of such an alert is shown in figure below.

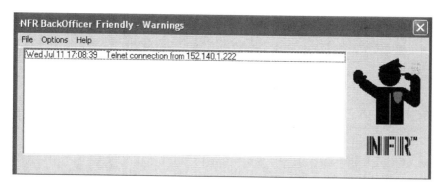

Figure 2.7 Telnet alert window.

Now let's try to establish a FTP connection. When we try to establish a FTP connection, it automatically gets closed by BackOfficier Friendly honeypot immediately after establishment as shown in figure below.

Figure 2.8 Closed FTP connection.

Simultaneously it generates an alert stating that a given I.P. address is trying to establish a FTP connection. An example of such an alert is shown in figure below

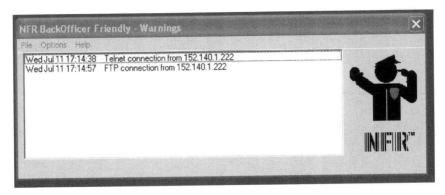

Figure 2.9 BackOfficer Alert.

Now we'll see how BOF works with selecting Fake Replies.

After selecting the Fake Replies option, when we try to establish a telnet connection, instead of closing the connection, BOF ask for login and password. This is shown in figure below.

Figure 2.10 Fake Reply by BackOfficer Friendly.

Figure 2.11 Fake Reply by making attacker enter password.

If either login or password, any one is wrong then it prompts up stating that login or password is incorrect and again asks for the login.

Figure 2.12 Further deception by BackOfficer Friendly.

And simultaneously generates an alert in BOF window. The alert so generated tells about the I.P. address which is trying to establish a telnet connection along with the user name and password of that user.

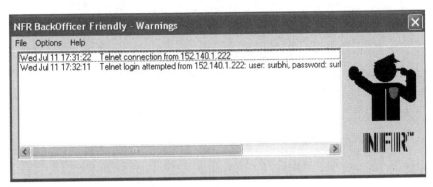

Figure 2.13 Alert showing attacker details.

If we don't enter the correct login and password within 1 min, connection gets terminated and an alert is generated in the BOF window about the login timeout.

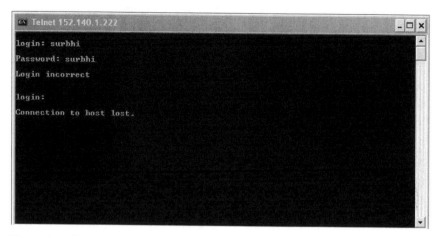

Figure 2.14 Terminated Connection.

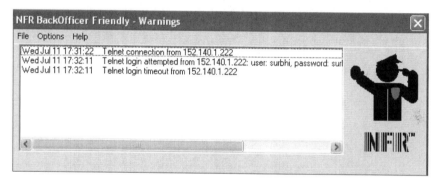

Figure 2.15 Alert about login timeout.

In case of FTP connection establishment after selecting Fake Replies, an error is generated stating that service is unavailable and then connection gets terminated.

Figure 2.16 Error state in case of FTP request.

Simultaneously an alert of such an attempt is generated on BOF window.

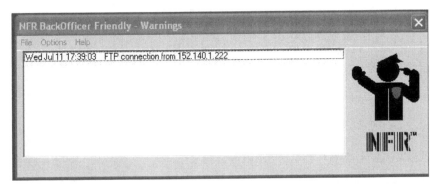

Figure 2.17 BOF window showing alert after FTP connection.

2.2 SPECTER

Specter [2] is a commercial production honeypot whose value lies in detection. It can completely simulate a machine. Thus has a capability to lure hackers away from the actual production machines. Specter can simulate 14 different operating systems in application level, which are most popular ones in both commercial and research world. These operating systems are:

1. Windows 98
2. Windows NT
3. Windows 2000
4. Windows XP
5. Linux
6. Aix
7. Solaris
8. MacOS
9. MacOS X
10. Tru64
11. NeXTStep
12. Unisys Unix
13. Iris
14. FreeBSD

Specter serves as good emulator of various inter services on these operating systems excluding Windows NT, Windows 2000 and Windows XP.

It is windows based software, which offers 14 different network services and traps like:

1. SMTP
2. FTP
3. TELNET
4. FINGER
5. POP3
6. IMAP4
7. HTTP
8. SSH
9. DNS
10. SUN-RPC
11. NETBUS
12. SUB-7
13. BO2K
14. GENERIC TRAP

These services although appear normal to the hackers but actually are traps, which persuade them to use their methods to falsely, use these services. Although hackers in their opinion are exploiting these services but actually they are leaving their traces to be monitored by the professionals of white hat community. These decoy systems appear to hackers as actual production system, which is performing all the activities of system whose activities they are emulating. But in reality they do nothing instead logs all the activities of the hackers for the appropriate guys of white hat community such as network administrator, etc. Specter is a low interaction honeypot. It is not capable to of utilizing the application to interact with the OS. It can just gather information about hackers such as Whois and DNS lookups, etc. It can just receive the hacker's request and fake the reply. It has no intelligence of its own.

Specter is associated with many default user names and logins well suited for English speaking environments, which can be easily cracked by hackers. It's not necessary to use these default user-names and logins, in fact custom user names and logins can be entered. This helps to further improve the integration of the Specter system into the network, moreover its quite beneficial if the common language in your organization is not English but something else. In fact you can use the names and logins of real people working for your organization. This helps to make the honeypot perfect, that is, it becomes difficult for hackers to differentiate between actual production system and Specter honeypot. Moreover you can even mix custom and default user names. Although Specter comes with massive amounts of decoy data, but you can add your own content to make the honeypot look even more like one of your production machines. For example, you can provide your own web documents for the simulated HTTP server. Specter can provide fake password files to intruders. It can send password files in 3 different format viz.:

1. Unix passwd
2. NT/W2K/XP binary uncompressed
3. NT/W2K/XP binary compressed

What kind of file is provided depends on the simulated operating system and the way an attacker tries to get the password file. The type of the password files can be specified. There are 7 types available:

1. **Easy**—Easy to guess passwords.
2. **Normal**—Normal passwords, some weaker, some stronger.
3. **Hard**—Hard to guess passwords.
4. **Mean**—Very strong, extremely hard to guess passwords.
5. **Fun**—Famous people as user names and appropriate passwords.
6. **Cheswick**—The famous password file from the well-known book "Firewalls and Internet security" by William R. Cheswick and Steven M. Bellovin.
7. **Warning**—The file sent is not a password file but a text file containing a configurable warning message.

It's very easy to install specter honeypot even on your personnel machine. The minimum requirements to install and enjoy the services of specter honeypot are:

1. Pentium III 800 MHz processor
2. 256 MB of RAM
3. Windows 2000 SP2 or Windows XP SP1
4. Display resolution of 1024x768

But the recommended requirements are:

1. Pentium 4 1700 MHz processor or better
2. 512 MB of RAM or more
3. Windows 2000 SP2 or Windows XP SP1
4. Display resolution of 1024x768

The main specter window looks like the one shown in the figure below.

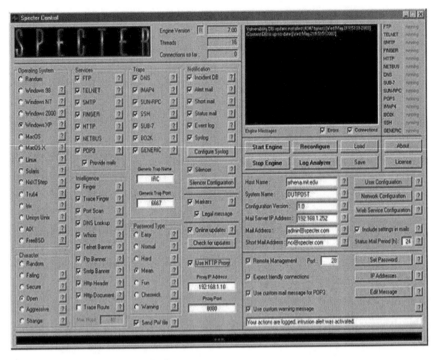

Figure 2.18 Main Specter Window.

Specter honeypot can work in 5 different modes viz.:

1. **Open**—The system behaves like a badly configured system in terms of security.

2. **Secure**—The system behaves like a well-configured system in terms of security.

3. **Failing**—The system behaves like a machine with various hardware and software problems.

4. **Strange**—The system behaves unpredictable and leaves the intruder wondering what's going on.

5. **Aggressive**—The system communicates as long as necessary to collect information about the attacker, then reveals its true identity by the appropriate means depending on the kind of connection and then ends communication. This is very handy to scare intruders away.

On selecting Specter system character mode as *"open"*, *"secure"* or *"failing"*, system appears to be suffering from various vulnerabilities. These vulnerabilities depend on the system used for simulating the services and services themselves. The most strong point about Specter to be noted over here is that these vulnerabilities keep on changing randomly. Thus it becomes virtually impossible to detect the presence of Specter honeypot based on combination of vulnerabilities and the services emulated.

Specter honeypot has power log analyzer capability. It can search through database to mine data corresponding to any particular incident and displays all the logged details about the requested incident.

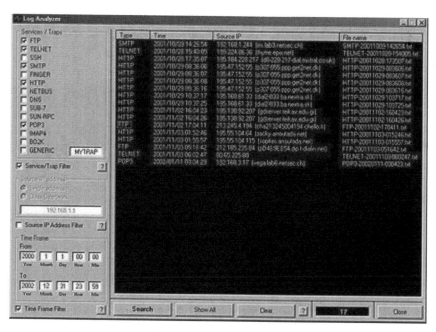

Figure 2.19 Details displayed by Specter.

Specter honeypot can also combine different available filters to search for a required incident. This is quite useful data mining capability available with Specter honeypot.

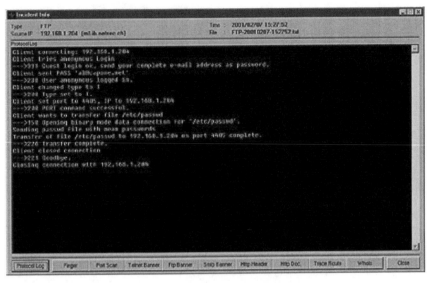

Figure 2.20 Data logged with a combination of filters.

All connections with Specter are logged with the IP address of the remote host, exact time, type of service and state of the honeypot machine at the time of the connection. Depending on the type of service/trap and the intelligence settings, there is much more information logged, such as a complete protocol log, content of mails sent, etc. Specter can log to its own incident database, to the system's event log or to a remote Syslog server. It can send alert mail messages right when something is happening. It can also send short alert mails suitable for redirection to mobile phones and pagers. Additionally, Specter can send status reports on a regular basis.

Specter offers various intelligence modules to collect information about attackers as well as evidence against them. The available intelligence options are:

1. FINGER
2. TRACER
3. PORTSCAN
4. TRACEROUTE
5. TELNET BANNER
6. FTP BANNER

7. SMTP BANNER
8. HTTP SERVER HEADER
9. HTTP DOCUMENT
10. WHOIS
11. DNS

Specter dynamically generates more than 100 different executable programs for various operating systems that will leave up to 32 hidden marks on an attacker's computer. The information contained in the marks can be valuable evidence in court.

2.3 MANTRAP

Another type of commercial honeypot is Mantrap [3]. Mantrap is also as popular as BackOfficer Friendly and Specter. It was produced by Recourse. This honeypot is implemented as a chrooted Solaris environment, designed to look and feel real to an attacker who gains access to it. Chroot (change root) is a unix mechanism that allows an administrator to force a process/process group to run under a subset of the file system, denying access to any other parts of the file system. It is possible for an hacker to guess that they are on a chrooted() ManTrap system by looking at the inode of the root directory (ls-id/). If it is high (usually within the 100000–200000 range), then the root directory is a chrooted() subset of a larger file system. This vulnerability, combined with hidden process disclosure (bugtraq ID 1908) should fairly accurately verify to a hacker (without root privs) that the host is a Mantrap honeypot, defeating its purpose. Recourse Technologies' Mantrap extends the honeypot into a full-blown deception host. The entire operating system—every service, file, and process—is presented to an attacker as a decoy environment. Mantrap automates many of the tasks necessary to perpetuate deception in its decoy environment. It populates the deception host with unique data, and can even update the data, to convince the attacker he's found a production host. Mantrap also attempts to identify the objectives of the intruder: is he trying to gain root, deface a web site, steal passwords, install Trojans, or what? Mantrap's analysis tools expedite audits of user/attacker activity.

This honeypot is categorized as mid-high level of involvement honeypot. It can be used as either a production honeypot (used both in detection and reaction) or a research honeypot to learn more about threats. As opposed to BackOfficer Friendly and Specter honeypots, mantrap does not emulate various services. Mantrap creates up to four sub-systems, often called cages or "jails". These jails are logically separated operating systems that are separated from a master operating system. Security administrators can modify these jails just as they normally would any operating system, to include installing applications of their choice, such as an Oracle database or Apache Web server. This makes the honeypot far more flexible, as the attacker has a full operating system to interact with, and a variety of applications to attack. All of this activity is then captured and recorded. This makes Mantrap more effective honeypot as using mantrap we can not only detect port scans and telnet logins but instead can also capture rootkits, application level attacks, IRC chat sessions, and a variety of other threats.

Although Mantrap is very useful as compared to other two above discussed honeypots but it has darker side also. Organizations have to understand these issues and mitigate these risks before deploying Mantrap systems.

Outbound access is a risk of any high-interaction honeypot. Mantrap cages are functional Solaris operating environments. These cages give attackers access to a wide range of capabilities. The more the attacker can do, the more you can learn. However, the more an attacker can do, the more damage he can potentially cause. Intruders are very good at identifying and breaking into vulnerable systems. If unfortunately hacker is able to compromise the Mantrap, then hackers can use its fully functional operating system to attack others. Since there are no mechanisms to stop attackers from using a compromised cage to launch denial of service attacks, scan other networks, or launch exploits against other vulnerable systems, thus once gaining control over the operating system everything goes according to their wish. Thus with Mantrap there comes the utter need to take suitable measures to mitigate this risk. Along with this, Mantrap is accompanied by other limitations also. The biggest one is that users are limited to what the vendor supplies them. Currently, Mantrap only exists on Solaris operating system.

2.4 HONEYD

Another most important category of commercially available honeypot is Honeyd [4]. Honeyd was created by Niels Provos in 2002. It is extremely powerful open source honeypot available software released under GNU General Public License. Currently Honeyd 1.5c has been released on 2007-05-27 and the next version is currently being developed.

Honeyd honeypot is mainly designed to work in Unix environment and it has the ability to emulate over 400 different operating systems and thousands of different computers. Like Specter, Honeyd emulates operating systems at the application level stack and it also emulates operating systems at the IP level stack.

Case 3

Honeyd in action:

"While trying to help the community to fight the evil worm MSBLAST, I looked at my favorite honeypot, called honeyd, to check if we could not play with the worm itself (Labrea played with another worm in the past... the past should not be forgotten). I restarted my honeyd.

Then, every hosts owned by msblast that was attacking my honeypot, first saw that the port 135 was opened. They then sent their evil payload on it, and naturally tried to launch their commands on the port 4444... I got the msblast.exe file in less than 5 minutes (internet is so evil you know...) and give it to a friend for an internal analysis in the rstack team."

Honeyd introduces several new concepts to honeypots. First, it does not detect attacks against its own IP address, as BOF and Specter do. Instead, Honeyd assumes the identity of any unused IP address. It then forwards the traffic of all non-existent systems to the Honeyd honeypot. Indeed Honeyd is receiving attacks by implementing ARP Spoofing. This layer2-based method binds an IP address of the intended victim (one which is currently not in use) to the MAC address of the honeypot. This way all systems on the network (including routers) send IP packets of non-existent system to the Honeyd honeypot. Any attempted connection to an unused IP address is assumed to be unauthorized or malicious activity. After all, if there is no system using that IP, why is someone or something attempting to connect to it?

Honeyd can monitor all of these unused IPs at the same time. Whenever a connection is attempted to one of them, Honeyd automatically assumes the identity of the unused IP addresses and then interacts with the attacker. This approach to detection has many advantages over traditional methods. Any time Honeyd generates an alert, you know it most likely is a real attack, not a false alarm. Instead of hammering you with 10,000 alerts a day, Honeyd may only generate 5 or 10. Furthermore, since Honeyd is not based on any advance algorithms, it is easy to set up and maintain. Lastly, it not only detects known attacks, but unknown ones as well. Anything that comes its way is detected, not only that old IIS attack, but also that new RPC 0-day attack no one knew about.

Honeyd can emulate different operating systems at the same time. As compared to Specter, it can emulate more than 13 different operating systems but it can only emulate one system at one time. Honeyd can emulate many different systems at the same time. It takes the very same database of signatures that Nmap uses and replies to Nmap probes based on the emulated operating system.

Honeyd is a small daemon that creates virtual hosts on a network. The hosts can be configured to run arbitrary services, and their personality can be adapted so that they appear to be running

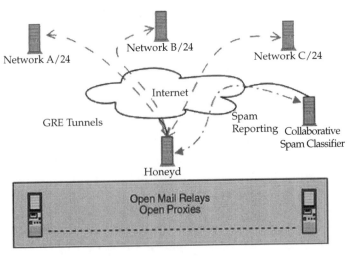

Figure 2.21 Honeyd Spam Research Overview.

certain operating systems. Honeyd enables a single host to claim multiple addresses on a LAN for network simulation. Honeyd improves cyber security by providing mechanisms for threat detection and assessment. It also deters adversaries by hiding real systems in the middle of virtual systems.

Honeyd can be used effectively to battle spam. Since June 2003, Honeyd has been deployed to instrument several networks with spam traps. We observe how spammers detect open mail relays and so forth. The figure above shows the overall architecture of the system. The networks are instrumented with open relays and open proxies. All spam emails are intercepted and analyzed for why they have been received. A single Honeyd machine is capable of simultaneously instrumenting several C-class networks. It simulates machines running mail servers, proxies and web servers. Captured email is sent to a collaborative spam filter that allows other users to avoid reading known spam. Moreover, this setup has also been very successful in identifying hosts infected with worms.

To implement Honeyd we need to compile and use two tools: Arpd and Honeyd. Honeyd cannot do everything alone and requires the help of Arpd. Arpd is used for ARP spoofing; this is what actually monitors the unused IP space and directs attacks to the Honeyd honeypot. Honeyd does not have the capability to direct attacks to it; it only has the capability to interact with attackers.

2.5 SUMMARY

This chapter discussed in details some of the commercially available honeypots. BackOfficer Friendly being the simplest is low interaction honeypot with very limited capabilities and a small set of services to emulate. Specter has more options and can generate fake replies, as discussed. Honeyd is the most popular Linux based honeypot that can simulate many machines at the same time. It is open source and most flexible of all low interaction honeypots. Mantrap based on Solaris is middle—high interaction honeypot and provides caged environment to the attackers while simulating multiple OS.

EXERCISES

1. Install the low interaction honeypot "BackOfficer Friendly". Monitor the network for 24 hours and review the alerts. Explain GUI interface as seen by remote control client of BackOfficer Friendly.

2. Discuss the value of following honeypots in detection, prevention and reaction to attacks:
   i) BOF
   ii) Specter
   iii) Mantrap
   iv) Honeyd

3. Discuss the use of nstat and nmap commands.

4. What are the risks associated with BOF, Specter, Mantrap and Honeyd. Discuss in details for each honeypot.

5. You have BOF. You want to implement the functionality of trap (as in Specter) using BOF. Is it possible? If yes, how? If no, why not?

6. Explain in details working, logging and information gathering in Specter.

7. What are cages in Mantrap?

8. What is Blackholing and ARP spoofing?

9. Explain the working of Honeyd.

10. Explain the working of Mantrap. How does Mantrap handles file system?

11. What is the advantage of using sniffer with Mantrap?

12. Contrast the following honeypots with respect to their advantages and disadvantages:
    i) BackOfficer Friendly
    ii) Specter
    iii) Mantrap
    iv) Honeyd

REFERENCES

[1] Back Orifice. http://www.cultdeadcow.com/tools/bo.html.
[2] Specter: www.specter.com/.
[3] Mantrap: www.recourse.com/products/mantrap/trap.html.
[4] Honeyd: www.honeyd.org/.

3

Honeynets

The honeypot solutions examined till now have progressed through increasing levels of honeypot interaction. This chapter concludes look at specific honeypot technologies with the most complex of all—the Honeynet. This chapter discusses value of Honeynets and presents Gen I Honeynets .Issues in Gen I Honeynets that lead to Gen II Honeynets are outlined. Further, Gen II Honeynets are discussed and the chapter concludes with sweetening Honeynets and looking at various risks associated with Honeynets.

3.1 OVERVIEW OF HONEYNETS

Honeynets [1] are high-interaction honeypots. In fact, it is difficult to conceive of a honeypot solution that can offer a greater level of interaction. The concept of a Honeynet is simple. It requires one to build a network of standard production systems. These network systems are placed behind some type of access control device (such as a firewall) and keep a track of the network. Attackers can probe, attack, and exploit any system within the Honeynet, giving them full operating systems and applications to interact with. No services are emulated, and no caged environments are created. The systems within a Honeynet can range from a Solaris server running an Oracle database to a Windows XP server running an IIS Web server, a Cisco router. In short, the systems within a Honeynet are true production systems.

Honeynets are a relatively new concept to the world of honeypots. Unlike honeypots, which have been in the public since 1990, development of Honeynets first began in 1999. Traditionally, almost all honeypot concepts were based on a single system configured to emulate other systems or vulnerabilities. A system was modified to create a jailed environment. A Honeynet is unique in that there are no modifications to the systems; they are identical to the builds found in an organization.

The concept of Honeynets developed over the next several years. In June 2000, the Honeynet Project [2] was formed. This is a nonprofit research group of 30 security professionals who volunteer their time and resources to researching attackers. These researchers cover a wide spectrum of the security community, including developers of IDS solutions, forensics experts, social psychologists, and intelligence officers. This diversity gives the group the capability to analyze and research a variety of aspects of the attackers. The organization's mission statement is "To research the tools, tactics, and motives of the attackers and share the lessons learned."

3.2 VALUE OF HONEYNETS

Honeynets [1] are an extremely flexible tool. They can fulfill any of the honeypot roles discussed thus far in this book. They can be used as a production honeypot, a resource that directly protects an organization. With respect to prevention, Honeynets excel at deceiving attackers, since they use authentic systems with real applications. When blackhats attack a Honeynet, they will have a difficult time determining that they are on a honeypot. Honeynets also can be used to detect attacks. As Honeynets use a variety of real systems, they can detect different attacks against unique systems. They also use the same model as Mantrap for detecting activity, by passively monitoring every port and every IP protocol. Honeynets are designed to capture not only the known but the unknown. They are also an excellent solution for responding to attacks. A Honeynet can run almost any conceivable operating system and application. This gives organizations the flexibility of using Honeynets for responding to system attacks. Although Honeynets easily adapt as production honeypots, they rarely are used that way because they are so complex. Honeynets require an extensive amount of time and resources to build, implement, and maintain. The effort involved with Honeynet technologies is often not worth the results for production honeypots. Yes, they excel at preventing, detecting, or reacting to attacks, but so do all the simpler solutions we have seen up to this point.

On the other hand, Honeynets add tremendous value as research honeypots. Their primary purpose in life and their reason for development are to learn about threats on the Internet: Who are the attackers? What the tools do they use? What the tactics do they employ? What motivates them? No other honeypot solution can obtain as much depth of information as a Honeynet. Within the field of research, Honeynets can add value in several areas.

3.2.1 Methods, Motives, and Evolving Tools

The first area of research is learning as much as possible about the attackers themselves. Honeynets can collect in-depth information about attackers, such as their keystrokes when they compromise a system, their chat sessions with fellow blackhats, or the tools they use to probe and exploit vulnerable systems. This data can provide incredible insight on the attackers themselves. The advantage with Honeynets is that they collect information based on the attackers' actions in the wild. You can see and learn step by step how they operate and why.

Case 1

The Honeynet Project used a Honeynet to capture a new tool and communication method never seen in the wild. In February 2002 a Red Hat server (a type of Linux) was compromised with TESO's wu-ftpd mass-rooter. The tool and method used to compromise the system within the Honeynet were not unique, and little was learned from the attack. However, what was unique was the backdoor the attacker put on the system. A backdoor is a mechanism that blackhats use to maintain control of a hacked box.

Traditionally such methods included installing a port listener on a high port or Trojaning the system binaries, such as /bin/login, to ensure the attacker always had remote access.

In this case, the attacker deployed a new tool. A binary was downloaded onto the hacked honeypot and then executed to run as a process. Following this, the Honeynet detected unique traffic going to and from the honeypot, specifically IP protocol 11, known as Network Voice Protocol. Most organizations would fail to pick up such a nonstandard IP protocol, since they traditionally monitor only the IP protocols TCP, UDP, and ICMP.

What many organizations forget is that there are many other IP protocols that can be used. However, the Honeynet quickly picked it up because it detects all activity.

3.2.2 Trend Analysis

As research honeypots [3], Honeynets also excel at trend analysis and statistical modeling. The information gathered can be used to predict attacks, acting as an early warning system. Traditionally, it has been very difficult to determine when an attacker was going to attack an organization or when a new tool has been released. Most detection methods are based on known signatures, such as Network Intrusion Detection Systems. These detection methods work by building a database of known attacks and then matching all network traffic against the database. Whenever there is a match, an alert is generated. Attempts have been made to use this information for trend analysis or to predict attacks. Organizations collect vast amounts of IDS alerts from various organizations, archive the data to a relational database, and then query the information for attack modeling. However, there are several challenges to this approach. The first is unknown attacks. Unknown attacks will not be detected or discovered. The second is false positives. Data from various organizations can have different levels of false positives, polluting the data and making statistical modeling far less effective. Honeynets are potentially a more effective tool for such research and prediction analysis. There are very few false positives with the data Honeynets capture. Almost all inbound traffic to Honeynet systems indicates some type of malicious activity. This makes the data far more reliable for analysis purposes. Even more important, Honeynets capture both known and unknown attacks. Honeynets do not use a database of known signatures for data analysis. They can discover attacks against new vulnerabilities just as easily as attacks against known weaknesses. By taking all the historical information a Honeynet collects and looking for specific patterns or changes in those patterns, researchers can determine new attack trends or predict future attacks.

Case 2

The Honeynet Project has demonstrated the capabilities based on 11 months of captured data during 2000 and 2001. Researchers identified specific behavior that could be used to predict attacks, such as intruders scanning a system or determining versions of specific services. In some cases, these behaviors provided up to three days' advanced warning that an attacker was going to launch an exploit.

Other analysis includes identifying new attack trends, such as a sudden rise in scanning for certain ports. This early indication and warning function is similar to the Navy's SOSUS systems. From the 1950s through the 1980s, enemy submarines posed a threat, since they could silently approach and attack from anywhere in the world's oceans. To detect these threats, devices were placed throughout the oceans' floor to passively capture the activity of enemy submarines. Honeynets can be considered the SOSUS of cyberspace, passively gathering information on threats. The data collected can then be used to analyze attackers' behavior and potentially predict new attacks.

3.2.3 Incident Response

A third area of research is incident response. A production honeypot can be used to help react to a specific incident or attack. A research honeypot, on the other hand, can be used to develop the general tools and skills used in any attack. Research honeypots provide a platform for organizations to develop and refine their response procedures in a controlled environment. This benefit is especially important, since organizations are reluctant to share information and facilitate public learning about actual attacks. Most organizations, when compromised, never publicly acknowledge the fact and may never contact law enforcement or other authorities.

Case 3

An FBI survey released in April 2002 found 90 percent of respondents detected computer security breaches in the past year, but only 34 percent reported those attacks to authorities. Instead, most organizations do everything they can to ensure the information is contained. Their logic is that bad publicity would be far more damaging than anything an attacker could do. This leaves the security community with very few public resources by which to learn about attacks or, more specifically, how to analyze them.

Contrast this situation with the U.S. Federal Aviation Authority (FAA). Whenever an American plane crashes, the incident is fully analyzed and shared with the entire aviation industry. The goal is to learn from the incident and ensure that it never happens again. Sadly, there is no such information sharing within the security community about actual attacks

---

However, Honeynets create a controlled environment for collecting information about attackers and imminent attacks. And agencies such as the Honeynet Project and the Honeynet Research Alliance are efforts to publicly share that information. Honeynets provide compromised systems for security professionals to analyze and learn from. The tools and skills developed can then be applied to real situations when production systems are attacked or compromised. In facilitating incident response, research Honeynets provide two major advantages. First, the information is not confidential, so it can easily be shared with the entire security community. For example, the images of the attacked system can be shared, giving professionals a real system on which they can practice and develop their response procedures. Individuals can use the same data set to share lessons learned with new tools and techniques. Such information dramatically improves the security community's ability to react to and analyze attacks. A second advantage that Honeynets provide with respect to incident response is the comprehensiveness of the information they collect. Honeynets capture all the attackers' activities on a real, complex system, from their keystrokes to every packet that enters or leaves the Honeynet. All of this captured data can be used as a kind of "answer book." Once an organization has analyzed a compromised honeypot using its standard tools, process and procedures, it can compare its results to the attacker's activities captured from the Honeynet—the "answer book."

This comparison can then be used to improve any incorrect analysis or failures in procedures.

3.2.4 Testbed

Finally, Honeynets can be used as a testbed, a controlled environment to analyze vulnerabilities in new applications, operating systems, or security mechanisms. When new technologies are deployed, they often have a variety of security issues. These issues can expose organizations to great risk. Before deploying such technologies, Honeynets can be used to test them in a highly controlled environment. By placing them in a Honeynet, the technologies can be monitored to determine if there are any risks or issues involved with them. For example, perhaps an organization wants to deploy new Web server functionality that ties in with a backend database. Such a technology could first be deployed in a Honeynet. Unknown vulnerabilities or risks can be discovered when blackhats attack the systems. There is little risk to the organization, since Honeynets control the activities of the attackers. The benefit is that the discovered risks can be addressed before the technologies are deployed in a production environment.

3.3 WORKING OF HONEYNET

Conceptually, Honeynets are simple mechanisms that work on the same principle as a honeypot. You create a resource that has little or no production traffic. Anything sent to the Honeynet is suspect, potentially a probe, scan, or even an attack. Anything sent from a Honeynet implies that it has been compromised—an attacker or tool is launching activity. However, Honeynets take the concept of honeypots one step further: Instead of a single system, a Honeynet is a physical network of multiple systems.

Honeynets are not a product you install or an appliance you drop on your network. Instead, Honeynets are an architecture that builds a highly controlled network, within which you can place any system or application you want. It is this architecture that is your Honeynet. The Honeynet operates as a kind of fishbowl, a self-contained environment in which you can see everything that happens. Also, like a fishbowl, in a Honeynet you can create any environment you want. In your fishbowl you can place different types of fish, stones, coral, plants, and lighting. In Honeynets you can place whatever systems and applications you want. Even though the systems placed within your Honeynet may be built identically to a production system, we define them as honeypots because their value within the Honeynet is being probed, attacked, or compromised. The captured activity within this controlled environment is what teaches us the tools, tactics, and motives of the blackhat community

There are three critical elements to Honeynet architecture: data control, data capture, and data collection. These elements define your Honeynet architecture. Of the three, the first two are the most important and apply to every Honeynet deployment. The third, data collection, only applies to organizations that deploy multiple Honeynets in a distributed environment. Data control is the controlling of the blackhat activity. Once a blackhat takes control of a honeypot within the Honeynet, his activity has to be contained so he cannot harm non-Honeynet systems. Data capture is the capturing of all the activity that occurs within the Honeynet. Data collection is the aggregation of all the data captured by multiple Honeynets. Honeynets are highly flexible: there is no specific way to implement a Honeynet solution. However, what is critical is that it meets the data requirements of Honeynet technologies.

3.3.1 Controlling Data

Data control is what mitigates risk. It controls the attacker's activity by limiting what can happen inbound and outbound. The risk is that once an attacker compromises a system within the Honeynet, she can use that system to attack other non-Honeynet systems, such as organizations on the Internet. The attacker must be controlled so she is not able to do that. It's fine if they attack other systems within the Honeynet, but we must protect non-Honeynet systems. One challenge to data control is automating it. Most often there is not enough time for manual intervention when governing a blackhat. When the attacker launches his exploits or Denial of Service attacks, it has to happen quickly enough to mitigate any damage. This means the response has to most likely be automated. Another challenge is ensuring that the attacker does not realize that his activity is being controlled.

There are eight requirements for data control. An organization can implement a Honeynet however they want, but the following functionality is critical to reducing risk:

i) Both automated and manual data control. In other words, data control can be implemented via an automated response or manual intervention.

ii) At least two layers of data control to protect against failure.

iii) The ability to maintain the state of all inbound and outbound connections.

iv) The ability to control any unauthorized activity. Unauthorized activity is defined by the policy of the Honeynet administrator. This implies some type of control to ensure that non-Honeynet systems are not harmed.

v) Data control enforcement must be must be configurable by the administrator at any time.

vi) Control connections in a manner as difficult as possible to be detected by attackers.

vii) At least two methods of alerting for activity, such as when honeypots are compromised.

viii) Remote administration of the data control. You must be able to remotely access and administer the data control mechanisms.

3.3.2 Capturing Data

Data capture is the second requirement for Honeynets. As with data control, a critical challenge is to capture all of the attackers' activity without them realizing they are within a Honeynet. Following are the requirements for effective data capture, as defined by the Honeynet Project:

- No Honeynet-captured data will be stored locally on the honeypot. (Data logged on honeypots is assumed to be unreliable and may be modified by intruders.) Honeynet-captured data is any logging or information capture associated with activity within a Honeynet environment.
- No data pollution can contaminate the Honeynet, invalidating data capture. Data pollution is any activity that is nonstandard to the environment. An example would be a nonblackhat testing a tool by attacking a honeypot.
- The following activity must be captured and archived for one year: network activity, system activity, application activity, and user activity.
- Administrators must have the ability to remotely view captured activity in real time.
- Captured data must be automatically archived for future analysis.
- A standardized log must be maintained for every honeypot deployed.
- Administrators must maintain a standardized, detailed writeup of every honeypot compromised.
- The Honeynet sensors' data capture must use the GMT time zone. Individual honeypots may use local time zones, but data will have to be later converted to GMT for analysis purposes.

Resources used to capture data must be secured against compromise to protect the integrity of the data.

3.3.3 Collecting Data

Data collection is the third of the three requirements. Data collection is unique in that it is not a requirement for standalone Honeynet deployments. Any organization deploying a single Honeynet most likely does not need this functionality. However, organizations deploying or managing multiple Honeynets most likely do need data collection. The purpose of data collection is to centrally capture and aggregate all the information multiple Honeynets collect. By correlating all of this collected information, organizations can increase the research value of their Honeynet deployments. There are four elements to data collection.

- Honeynet naming convention and mapping so that the type of site and a unique identifier is maintained for each honeypot. This implies some kind of IP/DNS mapping database.

- A means for transmitting this captured data from sensors to the collector in a secure fashion, ensuring the confidentiality, integrity, and authenticity of the data.

- An option for anonymizing the data. This does not mean to anonymize the data of the attacker but rather it gives the source organization the option of anonymizing their source IP addresses or other information they feel is confidential to their organization.

- Standardization on NTP, ensuring all data capture from the distributed Honeynet is properly synched.

3.4 HONEYNET ARCHITECTURES

We started off this chapter by stating that a Honeynet is not a prepackaged solution but architecture. The architecture is defined by three requirements: data control, data capture, and for multiple Honeynets, data collection. There are a variety of ways organizations can implement these architectures. We will cover two possible ways and discuss their advantages and disadvantages.

These two different architectures are known as GenI (first-generation) and GenII (second-generation) technologies. As its name implies, GenI was the first iteration of Honeynet solutions. Developed in 1999, GenIs were the first Honeynets to be deployed. While the GenI architecture accomplished its goals, a variety of improvements were identified, creating the GenII Honeynet, conceived in 2001. We will now look at how each technology works and its advantages and disadvantages.

3.4.1 Gen I

GenI technologies were first developed in 1999. Its primary purpose is to capture the activity of the blackhat community. There were already several solutions that had this capability, including most of the honeypot solutions that we have discussed so far. However, the GenI Honeynet has two advantages over most honeypot solutions: It can capture a great deal more information, and it can capture unknown attacks or techniques. GenIs were the first truly high-interaction honeypots.

GenI technologies [2] are limited in their ability to control and contain attackers, but they are extremely effective at capturing automated attacks or beginner-level attackers. They are primarily used to capture attacks that focus on targets of opportunity. GenI technologies are not effective at capturing advance blackhats, attackers that focus on targets of choice, for two reasons. First, GenI Honeynets are relatively easy to fingerprint; they have a variety of signatures specific to data control. Second, GenI technologies have little of value to attract advanced blackhats, being nothing more than the default installations of operating systems. This means most of the attacks GenI technologies capture may already be known. The architecture of a GenI Honeynet is simple. An isolated network is created that sits behind a network access control device, often a firewall as in Figure 3.1. Anything entering or leaving the Honeynet must go through the firewall. The Honeynet is on an isolated network to reduce risk. The goal is to ensure that it cannot attack any non-Honeynet systems. Often there is a separate production network for administration of the Honeynet and for collection of any data captured by the Honeynet. Additional devices can be added to the Honeynet architecture, such as routers, for additional control. The Honeynet itself then consists of various systems, each system in itself a honeypot. The goal is to capture and control any actiity to these honeypots.

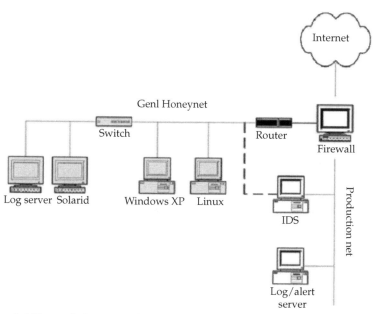

Figure 3.1 Network diagram of a GenI Honeynet.

The concept of data collection does not apply to GenI. For GenI Honeynets, we will focus on data control and data capture.

Data Control Issues and Methods

GenI Honeynets take a rather simple approach to data control, the purpose of which is to reduce risk and contain the outbound activity of compromised honeypots. Once a system is compromised, we must ensure that it cannot be used to harm other non-Honeynet systems, as defined in the requirements. This is done by creating a separate dedicated network for a Honeynet and then using a layer three, routing firewall as an access control device as in Figure 3.1. The firewall permits any inbound connections, but it controls outbound connections. The firewall contains the activity by counting every outbound connection from each honeypot. Anytime a honeypot attempts an outbound connection, the firewall counts the connection and keeps track. As the firewall is stateful, it is not counting packets but newly initiated connections. So an outbound FTP connection that downloads a 2-gigabyte file would be considered one connection. A compromised honeypot browsing to ten different Web pages would count as ten different connections.

The firewall is then given a certain limit of outbound connections. When this limit is met, the firewall blocks any further communication with the honeypot, effectively cutting off the attacker. The limit set by the firewall depends on the administrator; there is no right or wrong number for data control. The more connections a firewall lets outbound, the more activity permitted an attacker. This way, we can learn more, but the attacker can do more damage. The other extreme is to allow few or no outbound connections, greatly reducing the risk. One disadvantage to this approach is that the absence of an outbound connection can easily signify to the attacker that she is in a Honeynet environment. The attacker can then cause all sorts of malicious activity, such as deleting data or introducing false information. The Honeynet Project's first Honeynet deployment failed in this manner. The firewall used for data control allowed any inbound connection but blocked all outbound connections. An attacker exploited one of the honeypots but quickly discovered something was not right when he could not initiate an outbound connection. He then wiped all data from the hard drive and never returned. How many connections you allow outbound depends on what you are attempting to learn and how much risk you are willing to assume. If you want to capture automated attacks, such as worms, you can block all outbound connection or perhaps allow only one or two. This would allow the worm to gain control of a honeypot, initiate an outbound connection to download any instructions or payload they require, but block any further scanning attempts. There is no concern about signature detection, since there is no live human to detect anything. For human attackers, we most likely have to allow a certain number of outbound connections. If we block all outbound attempts, we risk detection. Also, we cannot research the behavior of the attacker if he cannot do anything. So we have to allow outbound activity.

However, we also have to limit this activity. What happens if the attacker uses the honeypot to scan millions of other systems on the Internet or perhaps launches a Denial of Service attack (which happens far more often than you think). A compromise has to be met. In general, members of the Honeynet Project have found that allowing five to ten outbound connections per day works best. This number gives attackers flexibility, such as initiating outbound connections to download toolkits or establishing Internet Relay Chats for communications. However, it is limited enough to block most attacks, such as Denial of Service attacks or scanning. To implement limited outbound connections, a firewall solution requires the capability to count connections and respond when the limit has been met. There is no right or wrong way to do this.

Here we have a data control rulebase for a CheckPoint FW-1 Next Generation firewall. This rulebase is designed for a GenI Honeynet deployment. It allows anything inbound to the Honeynet but counts the outbound connections and blocks when the limit is met.

Rule 1 specifies which systems can make an administration connection to the firewall—in this case the system fw-admin. This is a common best-practices rule for most firewall deployments.

Rule 2 blocks any other connections to the firewall, including any scans or attacks. This is a common best-practices rule for most firewall deployments.

Rule 3 allows any inbound connections to the Honeynet except anything from our production network. Remember, the production network should never directly communicate with honeypots within the Honeynet. This would cause data pollution and potentially give away the production network's identity to attackers. Notice that the tracking action is Mail. This means an e-mail alert will be generated for each inbound connection.

Rule 4 allows compromised honeypots to initiate an outbound connection to any systems except the production network. We do not want our attackers within our production systems.

Rule 5 denies all other activity that does not match any of the previous four rules. This is a common best-practices rule for most firewall deployments.

One of the requirements for data control is to have two layers of control. This redundancy ensures that there is no single point of failure. The two layers also give organizations flexibility in containing an attacker's activity. With GenI Honeynets, the second layer is a router placed in between the firewall and the honeypots as in Figure 3.1. The purpose of the router is to both screen the firewall from attackers and act as a second data-control mechanism. The router screens the firewall by preventing attackers from seeing the firewall. Once a honeypot is compromised, the bad guys will attempt to make outbound connections. However, they will not see the firewall that is controlling them but a router, which is most likely what they expect to see. As a second means of data control, the router can be used for egress filtering, ensuring no outbound spoofing attacks are launched. This effectively blocks the most common Denial of Service attacks. It can also be used to block commonly used ports for scanning or attacks, such as portmapper. The router can also block traffic that is difficult to maintain state on, such as ICMP. The combination of a layer three firewall and router creates an effective, automated mechanism for data control. Another requirement for data control is alerting. We have to be alerted whenever there has been a violation of data control requirements. For example, we want to receive an alert whenever an inbound connection is attempted. This indicates a probe, scan, or perhaps even an attack. An even higher priority is receiving an alert whenever a honeypot attempts an outbound connection. This indicates a honeypot has been compromised and an attacker or an automated tool is attempting an outbound connection, perhaps to receive a toolkit. Such alerting is normally done by the firewall, the same mechanism that is tracking all inbound and outbound connections. For example, with the FW-1 solution, the same script that tracks and blocks inbound and outbound connection also generates and sends alerts for all activity. The alert, generated by the firewall, warns that a honeypot has initiated an outbound connection and the system is most likely compromised.

Data Capture Issues and Methods

The second requirement for Honeynet deployments is data capture. The challenge is to capture as much data as possible without the blackhat knowing his every action is captured. This is done with as few modifications as possible to the honeypots. Any modifications are potential fingerprints for the attackers to detect, and they can potentially pollute the data collected. Also, captured data cannot be stored locally on the honeypot. Information stored locally can potentially be detected by the blackhat, alerting him that the system is a Honeynet. The stored data can also be lost or destroyed. The key to these challenges is capturing data in layers. You cannot depend on a single layer for information. You gather data from a variety of resources, reducing the risk of failure while increasing the information gathered.

The first layer of logging activity is the firewall. Previously, we discussed how we can use the firewall to control data. This same firewall can be used to capture data. The firewall makes an excellent data capture layer, since all traffic must flow through it without the attacker's knowledge. The information a firewall can log is limited, similar to a low-interaction honeypot. It cannot capture the attacker's keystrokes, nor can it capture packet payloads. Instead, a firewall logs primarily packet header information, such as the date/time of the attack, the source and destination IP address, and the source and destination port. However, as with low-interaction honeypots, such information can be extremely useful, especially for trend analysis and statistical modeling. Also, since we are not using a database of known signatures, firewall logs are excellent for detecting new attacks or changing trends in scanning.

The second critical layer is the IDS system. As you can see in Figure 3.1, an Intrusion Detection System is deployed on the Honeynet. The sensor has two interfaces: one connected to the Honeynet and one connected to the production network. Normally, such a dual-homed device is a very risky idea, since it gives an attacker a direct connection between two networks. However, in this case the interface on the IDS sensor connected to the Honeynet does not have an IP address associated with it, so there is nothing for attackers to attack. Instead, the interface (designated by the dotted line) is a passive interface that captures all activity on the network. The second interface connection to the production network allows remote administration and data collection from the sensor.

The first most important role of the IDS sensor is to capture all network activity. It is used to capture and record every packet and its payload that hits the wire. Most organizations cannot afford to aggressively capture so much information; their resources would simply be too overwhelmed. However, Honeynets have very little activity. What they do have is most likely some form of probe or attack. As such, they can afford to capture all the packet activity. This information is often the most critical, since it allows the Honeynet to not only analyze attacks at the network level but to capture keystrokes, toolkits, and even communications between blackhats. This captured information is most often stored by the IDS sensor to a binary log file, which is then later reviewed by an analyst. Once again, with the IP protocol 11 packets we discussed earlier, the communications were captured and then decoded from the IDS capture of all network activity.

The second function of the IDS system is to alert to any suspicious activity. Most IDS sensors have a database of signatures that represent known probes or attacks. When a packet on the network matches one of these signatures, the sensor generates an alert. For most organizations, alerting on signatures is the primary purpose of an IDS deployment. For Honeynets this information is not critical. All inbound and outbound activity is by default considered suspicious or malicious, so the IDS alerts have limited value. The firewall is already alerting us on inbound and outbound activity. Also, these alerts work only on known attacks; unknown attacks for which there are no signature will not be alerted. Even if attacks are known, the signature database has to be updated for new discoveries.

However, IDS systems can give detailed information about a specific connection. This additional information can tell us what an attacker is doing. One possible solution for deploying an IDS sensor is Snort, an Open Source solution. Snort can easily be configured to capture all network activity to a binary log file and alert to suspicious activity. The binary log files are the critical information, since they capture all the network activity. In addition, Snort can be configured to log to a separate file all ASCII communication (such as keystrokes from an FTP session) captured at the network level. The alerting mechanism of Snort is not critical but is used for an additional source of information.

A third layer for data capture is the honeypot systems themselves. We want to capture—both locally and to a remote log server—all system and user activity that occurs on the honeypot. The log server is a remote repository for all the system logs. Even if the local logs on a honeypot are modified or destroyed, there is still a second copy for the deployment of the log server within the Honeynet.

Within the Honeynet we do not want to make any attempt to hide the use of a remote syslog server. If the blackhat detects this, the worst the attacker can do is disable syslogd (which is standard behavior for most blackhats). This means we will no longer have continued logs, but we will at least have information on how the attacker gained access and from where. Once they detect the remote log server, more advanced blackhats will attempt to cover their tracks by compromising the system. They want to ensure that they leave no traces of their activity. Standard procedure is for them to wipe or modify the system logs, hiding their actions. If a remote log server is used to store duplicate copies of system logs, advanced attackers may also go after the log server to hide the copied logs of their activity. These attacks against the log server are exactly what we want to see. The syslog server is normally a far more secure system. For a blackhat to successfully take control of such a system he will have to use more advanced techniques, which we will capture and learn from. If the syslog server is compromised, we have lost nothing. Yes, the blackhat can gain control of the system and wipe the logs. But don't forget: Our IDS system passively captures and records all of the logging activity that happened on the network. In reality, the IDS system acts as a second remote log system. A second method for capturing system data is to modify the system to capture keystrokes and screen shots and remotely forward that data.

Limitations

There are several problems with GenI technologies, specifically with data control. The first is risk. If we allow only ten outbound connections, the attacker can successfully launch ten exploits. After the tenth connection, all future attempts are blocked. However, the attacker can leverage the flexibility we give him. For example, once the attacker gains access to a honeypot, he may make an outbound connection to download his toolkit. When connection is launched, the tool can attack nine systems before we reach our limit of ten connections. With GenI, risk is reduced as mass scanning, rooting, or Denial of Services are designed not to happen. However, attackers can still do damage with the first number of allowed connections.

The second disadvantage is fingerprinting. Once a bad guy compromises a system within a Honeynet, he can potentially launch outbound connections and see if a limit is met and if any future connection is blocked. If all of a sudden the attacker's outbound activity is blocked, she may be able to fingerprint the Honeynet. Also, the use of a layer three firewall is easy to identify, since all traffic that passes through the firewall has TTL (time to live) decrement. Traditionally, fingerprinting has not been a problem.

The last disadvantage with GenI is in data capture. Traditionally, keystrokes and user activity were captured at the network level. IDS sensors such as Snort could capture protocols such as FTP, Telnet, or HTTP. These protocols are cleartext, so we could monitor the attackers' connections by capturing the keystrokes at the network, this option no longer works, since the bad guys are now also using encryption. As such, more advanced methods must be used. Trojaned shells have worked in the past, but they are limited to Unix systems.

All of these issues—data control and data capture—are addressed with GenII.

3.4.2 Gen II

GenII Honeynets, developed in 2002, were created to address a variety of the problems found in GenI technologies. Specifically, they are easier to deploy and harder to detect. The vast majority of changes were made in how data control is done. Like a GenI Honeynet, a GenII Honeynet [4] is neither a single system nor a software solution. GenII Honeynets are architecture designed to capture and analyze threats on the Internet. However, this architecture is radically different from GenI.

Data Control Issues and Methods

GenI technologies perform data control with a firewall that counts and limits the number of outbound connections. This solution, while effective, is not very flexible and is easy to fingerprint. GenII Honeynets as shown in Figure 3.2 address the problem by modifying the Honeynet architecture.

The first difference is the use of a single Honeynet sensor. The Honeynet sensor combines the functionality of both IDS sensor and the firewall seen in GenI Instead of having to deploy several devices, we only have one. This makes it much easier to deploy and manage. The biggest difference is the use of a layer two firewall, combining both IDS and firewalling functionality.

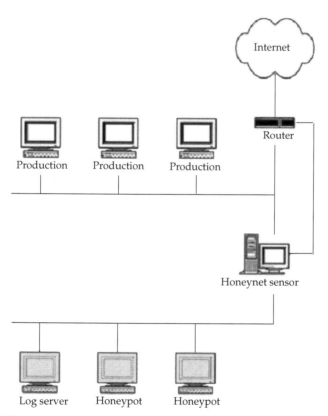

Figure 3.2 Network diagram of a GenII Honeynet.

The second major difference is in the Honeynet sensor itself. The sensor is a layer two device similar to a bridge. This makes the device much harder to detect. Unlike the layer three firewall in GenI technologies, with the Honeynet sensor there is no routing of packets, no TTL decrement of system hops, and no MAC device numbers for attackers to detect. Several operating systems support the capability to operate as a bridge, including Linux and OpenBSD. The device is nearly invisible to any attacker, making it much harder to detect. However, like GenI technologies, every packet entering or leaving the Honeynet must go through the Honeynet sensor. Because they use a layer two device, GenII Honeynet deployments can be part of a production network instead of being on an isolated network such as GenI Honeynets or other honeypots. In Figure 3.2, the layer two Honeynet sensor divides production systems from the Honeynet system, while in reality all the systems are part of the same network. The separation is happening at layer two, as opposed to layer three. Besides the architectural changes, data control mechanisms in GenII Honeynets are radically different from those in GenI. Instead of relying on a layer three firewall that applies access controls based on IP headers, GenII uses a technology called IDS gateway. An IDS gateway is similar to a firewall in that it controls who can access what resources on the network. However, an IDS gateway not only blocks connections based on the service, but it also has the intelligence to distinguish between an attack and legitimate activity. The IDS gateway has a signature database built into it. When a known attack is matched against the database, the connection can be blocked. This technology combines the functionality of firewalls with the capabilities of Intrusion Detection Systems. For example, most stateful firewalls would allow any HTTP connection to the Web server, regardless of the connection type. As a stateful firewall, it is only analyzing the IP headers and port information. If an attacker launched a Web attack, such as the attacks CodeRed II launched, the firewall would allow the connection through. However, IDS gateways have the ability to detect the attack by matching it against its database of known signatures. Even though the system is attempting a connection to HTTP, this specific connection will be blocked because the gateway can determine it is a known attack, specifically CodeRed II.

This technology has several advantages when used for data control. The first advantage with an IDS gateway is the ability to detect unauthorized activity. Instead of tracking the attacker's [3] activity by counting the number of outbound connections, we add more intelligence by tracking what his activity is. We will identify unauthorized activity by his actual actions and intent. If an attacker attempts ten outbound FTP connections, that will be fine. However, if he attempts a single outbound FTP exploit against a non-Honeynet system, then that activity must be contained even if that attack is within the ten-connection limit. This IDS gateway technology does have limitations. It can only detect attacks with known signatures, so unknown attacks with no signature can bypass this technology. That is why this technology is usually combined with GenI concepts as a backup mechanism.

The Honeynet sensor may also count outbound connections, but it has far greater threshold, such as 50 outbound connections. This technology can help reduce risk and make fingerprinting much more difficult, as compared to GenI Honeynet.

The second advantage is in the way GenII technologies respond to unauthorized activity. Instead of simply blocking connections, they modify or throttle the attacker's activity. These responses will be far more difficult for the attacker to detect. This is accomplished by modifying packets as they travel through the layer two gateway. For example, once an attacker has taken over a system within the Honeynet, he may attempt to launch an FTP exploit against a non-Honeynet system. With GenI technology, the data control is limited; after the tenth attempt outbound, all further activity, including any exploits, would be blocked. However, with GenII technology, the exploit attempt would be identified and then modified to make the attack ineffective. The layer two gateway would modify several bytes within the exploit code, disabling its functionality, and then allow the crippled attack to proceed. The attacker would see the attack launched and packets return but would not understand why his exploit never worked. This approach allows us to gain better control of the attacker's actions without his knowledge. Another example would be throttling unauthorized activity. An attacker may attempt to scan or launch Denial of Service attacks against non-Honeynet systems. These attacks would pass through the layer two gateway, which has the ability to drop certain packets or entire connections. It also has the ability to generate fake responses, such as blocking entire connections, but returning RST packets to the attacker, thus forging a dropped connection. Once again, GenII technologies have more flexible responses that are harder to detect. One example of IDS gateway technologies is Hogwash. Hogwash is an Open Source solution that modifies the Open Source IDS Snort. The modifications give Hogwash the ability to drop and modify any packet that passes through the gateway.

Data Capture Issues and Methods

GenII technologies use many of the same mechanisms from GenI for data capture. A large percentage of captured information comes both from the firewall logs and the network data captures from the sniffer. In fact, in many cases the network data captures still proves to be the most critical information when examining an attack. However, for capturing the attacker's activities, especially keystrokes, new mechanisms must be developed. This information must be captured from the actual honeypot. Most attackers today, even the script kiddies, use encryption, such as SSH, to communicate with hacked systems. To capture the keystrokes, we need to capture them from the Host system. In GenI technologies, the approach was to make modifications to the system binaries, such as /bin/bash, to capture the data. With GenII, there are more advanced methods that are more reliable and harder to detect. Kernel modules are currently being developed that modify the system kernel to record the attacker's actions, similar to what we saw with ManTrap in the previous chapter.

Data Collection Issues and Methods

GenII Honeynets are designed to be deployed in a distributed environment where multiple Honeynets are controlled by a single organization. For such deployments, data collection becomes a critical issue. For a single Honeynet deployment, data collection happens at the Honeynet itself or on the administration network. However, for distributed deployments, there must be a means for remotely managing the systems and collecting the captured data.

The data collection requirements for GenII Honeynets define how data can be collected from distributed Honeynets and in what format. These standards ensure that multiple Honeynets can work together and share their findings.

The most critical aspect of data collection is ensuring that the information is collected in a secure fashion. The central location point for collecting data has to guarantee the integrity, authenticity, and confidentiality of the information. This implies some type of encryption, such as IPsec tunnels from each distributed Honeynet to the central location point. The encryption ensures that the data sent has not been tampered with in transit, that each Honeynet authenticates itself to the central server, and that no other third party has seen the data.

In Figure 3.2 there is a third interface on the Honeynet sensor, specifically a solid line going to the router. This third interface is dedicated to data collection and remote management. Unlike the other two interfaces, this is a layer three interface with an active IP stack. This allows the gateway to communicate with other systems not on the network.

Over this connection, all data is remotely sent to a central collection point, and all distributed Honeynets are managed.

The other critical element of data collection is standardizing the format of the data sent. This ensures that data collected from different Honeynets, and potentially different organizations, can easily be shared and aggregated.

GenI technologies are designed for capturing automated attacks or blackhats that attack targets of opportunity. GenII technologies are designed to capture more advanced blackhats, those who focus on targets of choice.

3.5 SWEETENING THE HONEYNETS

A Honeynet consisting of default installations of systems and little or no activity is a sterile environment. Such an environment is excellent at capturing automated activity or attackers focusing on targets of opportunity. However, such a sterile environment has little value to advanced blackhats, individuals who concentrate on targets of choice, high-value systems. For such attackers, one must sweeten the pot. There are a variety of steps that can enhance the value of a Honeynet to create the illusion of a high-value target. We will cover just a few of the possible ideas.

The first thing one can do is to assign your Honeynet a valid domain name. Register a domain name you feel will attract the clientele that you want to research. Then build a Honeynet along this theme. For example, you may be interested in studying the tools and tactics used by blackhats concentrating on e-commerce sites. You can create and register an e-commerce name, such as "Making Money" or "Finances-R-Us" and then emulate an e-commerce site based on the domain name, or you could create an official Web site that extols the virtues of your e-commerce site, but explains that a large part of it is still under construction. Blackhats will potentially find the site by digging through Internet databases or online search engines, just as they would against valid targets.

There are also ways to create what appears to be valid activity within your Honeynet. You can create multiple users on your systems, including full names, home accounts, and even mail. Register these accounts to various mail lists so they periodically receive e-mail. Automated jobs can be created to add activity to the Honeynet.

Another idea is to create e-mails or documents with bogus information, such as false user accounts and passwords. You can monitor blackhats and see if they read these e-mails or documents. How much social engineering you add to your Honeynet environment depends on who your target is and what you want to learn. The more value you create with your Honeynet, the more you can potentially learn about advance blackhats, their tools, and their techniques.

3.6 RISKS ASSOCIATED WITH HONEYNETS

Without a doubt, Honeynets are the riskiest of all honeypot solutions. Organizations must be fully aware of these issues before deploying Honeynet technologies. There are two primary reasons for this high level of risk. The first is the level of interaction. Attackers are given complete access to full operating systems. Once they have access, there are no limitations to what they can do to the system. Attackers can use the Honeynet systems to compile code, launch attacks, or distribute tools. The only thing limiting the attacker's activities is the data control mechanism on the outside of the honeypots.

The second reason for the tremendous level of risk is the complexity involved. A variety of technologies must work together. Firewall rulebases have to be properly configured and maintain the state of inbound and outbound connections. Alerting scripts must activate and archive information on unauthorized activity. System logs must capture system and user activity and forward that to remote log servers. A failure in any one of these configurations or technologies could expose a Honeynet to great risk. For example, a misconfigured script could fail to block outbound connections, allowing an attacker to launch attacks against other systems. Or perhaps a bad signature in the IDS signature database will cause the IDS process to die. The more dependencies a technology has, the greater the chance of something failing.

A third reason Honeynets are high risk is that they are designed to capture and control certain expected activity. A new and unexpected threat can bypass the security mechanisms of Honeynets. Perhaps a blackhat develops a new tool or tactic, one that can bypass the data control mechanisms of a Honeynet, allowing him to attack other systems. An example of such technology is ADMutate, developed by K2. It uses polymorphic shellcode to obscure common attacks, hiding them from IDS sensors. Another threat is if the blackhats develop a method to bypass all data capture functionality, leaving us blind to the attacker's activities. Organizations must seriously consider the risk of Honeynet technologies before deciding to deploy one.

3.7 SUMMARY

Honeynets are a high-interaction honeypot. In fact, they are most likely the most high interaction solution possible. A Honeynet is not a commercial solution you can buy or install but an architecture that creates a highly controlled environment. It is flexible, has extensive data capture capabilities and is adaptable to many organizations.

EXERCISES

Short Answer type Questions:

1. What are the gaps in Gen I Honeynet that led to design of Gen II Honeynets?
2. Explain the architecture of Gen II Honeynet.
3. What do you meet by sweetening the Honeynet?

Long Answer type Questions:

1. Explain in detail the working of Honeynets.
2. Discuss the architecture of Gen I Honeynets.
3. Discuss various issues in Gen I Honeynets.
4. How can you sweeten a Honeynet?
5. Enumerate and explain the various risks associated with Honeynets.

REFERENCES

[1] "Know Your Enemy: Honeynets," *The Honeynet Project.* http://www.honeynet.org, 2006.
[2]   Philippine Honeynet Project. http://www.philippinehoneynet.org.
[3] L. Spitzner, "Honeypots, tracking the hackers" http://www.tracking-hackers.com 2002.
[4] "Know Your Enemy: GenII Honeynets," *The Honeynet Project.* http://www.honeynet.org, 2005.

4

Attacks and Role of Honeypots

After having a sound basis of honeypots and knowledge of commercially available honeypots, this chapter further delves into details about how honeypots provide a line of defense in each and every phase of attack. It tells the essence of honeypots and the reasons to use them for prevention, detection and reaction to attacks and in research. It further discusses the role of honeypots in defending against major and costly attacks on the Internet like worms, viruses, spam, phishing mails and distributed denial of service attacks.

Focusing on the value of honeypots, they can be subdivided into two general categories viz. production purposes or research purposes. Production purposes honeypots are used for protecting an organization. Their activities include preventing, detecting or helping organizations to respond to an attack. Research purposes honeypots are used for collecting information. Research honeypots are used for different purposes by different organizations. The information collected by them is of different value to different organizations. Some use them to study trends of hacker activity and some are just interested in early warning and prediction of attack, or law enforcement. In general we can say that low interaction honeypots are usually used for production purposes and high interaction honeypots are used for research purposes. However, either type of honeypots can be used for either purpose. While protecting an organization (production honeypot), honeypots may work in one of three ways; prevention, detection or response.

4.1 PHASES OF ATTACK AND HONEYPOTS

According to security goals, honeypots can be divided into four broad categories, namely prevention, detection, reaction or response and in research [1].

Prevention honeypot stops attacker compromising production system indirectly. Hacker wastes time attacking honeypot system instead of production system.

Detection honeypot gives alert when attack occurs. It detects compromises by analyzing system activities and is effective in detecting new or unknown attacks. It reduces both false positive rate and false negative rate.

Reaction honeypot provides an environment similar to production system for taking measures to find the cause and patch vulnerabilities after the production system is attacked and compromised. Taking a production system off-line for a full analysis is not always feasible and may be a loss after intrusion occurs. Reaction honeypot removes the difficulty.

The goal of research honeypot is security research. Researchers analyze new attacking tools as well as worms extracted from recording information. Remedies or solutions can be applied to enhance normal system security.

4.1.1 Prevention

Honeypots can prevent hacker from attacking in several ways:

1. Protection against attacks based on tools: The attack tools randomly scan the entire network and search for vulnerable systems. On finding vulnerable systems, these automated tools attack that system and take over it, like worms self-replicate them and copy themselves to the victim. To protect systems having vulnerabilities against such attacks, honeypots try to slow down the scanning activity of attack tools or they even sometimes stop the tools from scanning. Such honeypots are known as sticky honeypots. They uninterruptedly keep monitoring the unused IP space. When a scanning activity by

an attack tool comes in their notice, they interact with it and slow it down and sometimes are even successful in completely stopping them down. Sticky honeypots use various TCP tricks, such as a reducing TCP windows size of zero, so the no further message is transmitted, putting the attacker into a holding pattern etc. for doing so. These tricks are quite beneficial for slowing down or preventing the spread of a worm that has penetrated an internal organization. LaBrea Tarpit is an example of such a sticky honeypot.

2. Protection against human attackers: Deception and deterrence concept is used to protect against human attackers (Discussed in details in Chapter 5). The aim is to confuse the attacker. The attackers' time and resources are wasted in interacting with honeypots. While the interacts with honeypots, the organization, which has detected an attack, gets an ample amount to time to respond and stop the hacker. This is a deception concept. In deterrence concept, the attacker comes to know that the organization is using honeypots. But he does not know which systems are legitimate computers and which are honeypots. Thus due to fear of getting caught, the attackers do not attack the organization. Thus honeypots have deterred an attack on the organization. Deception Toolkit is a good example of such a honeypot.

4.1.2 Detection

Production honeypots can be used for attack *detection*. Even though an organization is made highly secure and equipped with the latest security tools, then also failures may occur. Attackers are always one step ahead of defenders. Failures may even occur due to human mistakes. Thus detection is highly critical for detecting an attack so that one can quickly react to them, and stop or mitigate the damage they do to an organization. Detection techniques like IDS sensors and system logs, etc., used earlier have become quite ineffective for various attack scenarios. They generate huge amount of data, large percentage of false positives and are unable to detect new attacks. They are also unable to work in encrypted or IPv6 environments. Honeypots generate only small and useful data sets, thus reduce false positives. They are also successful in capturing unknown attacks such as new exploits or polymorphic shellcode and can work in encrypted as well as IPv6 environments.

4.1.3 Response

Response is the most challenging line of defense provided by production honeypots. Normally even after detecting an attack, we have very little information about who the hacker is, how he got into the system and how much damage he has done till now to the system. But to respond or react to an attack, such information is needed in detail. In such situation we often face two problems. Firstly, not all the attacked systems can be taken offline for analyzing. So organization is limited to analyze the compromised system live, while it still provides the services. Thus we can't analyze the system for what had happened, how much damage has been done to it, etc., to a maximum extend. Secondly, even if we are successful in pulling down the system offline, we are not able to determine what damage has been done by the hacker as data is already polluted to a large extend as there as already been so much activities like user log in, mail accounts read, files written to database etc. Thus we are not able to determine what normal day-to-day activity is, and what hacker's is. Honeypots are quiet useful in such situations. They are excellent response tools and can address both of these problems. Honeypots can be very easily taken offline for full analyses. Pulling them offline does not cause any harm to normal day-to-day activities. Also since only unauthorized and malicious activities are captured by honeypots, they are very good tool for addressing second problem also. Honeypots are much easier and efficient for analyses than hacked system because any data retrieved from honeypots is just related to hacker only. Thus honeypots are quiet a lot valuable here. They can provide you with in-depth information needed to rapidly and effectively respond to an attack.

4.1.4 Research

Other than being powerful tools for production purposes, honeypots are also powerful tools for research purposes. Along with protecting your organization, honeypots can also be used to gain extensive information on threats that only handful of other technologies is capable of gathering. Like military organization that depend on information about who their enemy is, to better defend against them, information security can also defend itself in a much better way if it gets to know about who its enemy is. Thus at present the greatest problem faced by our security professionals is a lack of information or intelligence on cyber threats. These problems are addressed by research honeypots. They are used to collect information on threats. This information can be used for trend analysis, identifying new tools or methods, identifying hackers and their communities, early warning and prediction, or motivations and other various types of purpose.

4.2 WORMS AND VIRUS ATTACKS

4.2.1 Worms

This section evaluates the usefulness of honeypots to fight Internet worms [2]. The worms have become ubiquitous and exist in the networks, consuming network and system resources. A honeypot framework can be used to fight off Internet worms and even perform a counterattack.

An Internet worm is a mischievous code that spreads itself over networks [3]. It attacks vulnerable hosts, infects them, and then uses them as means to bounce or propagate to other vulnerable targets. While virus infections are induced due to the problems of abusing human vulnerabilities through social engineering, such as encouraging a user to click on an email attachment, Internet worms usually are due to technical vulnerabilities. Most of the time, worms are written and developed by computer hackers, researchers in the security field and virus authors.

From the functional standpoint [4], there are three primary actions of an Internet worm:

1. Infection: infect a target by exploiting vulnerability.
2. Payload: launch malicious actions on the local infected target or toward others remote hosts.
3. Propagation: use the infected target as means for external propagation.

Figure 4.1 shows the lifecycle of a worm from its birth to death.

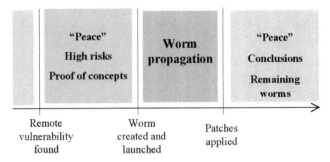

Figure 4.1 Birth and Death of Worm.

Honeypots are found to be very advantageous to fight off Internet worms. Honeypots can be used as a defense during all the three different phases of a worm i.e., infection, payload, and propagation.

Role of Honeypots during Worm Infection Phase

The infection phase of a worm is when it abuses a vulnerability to copy itself onto a chosen target. During this phase, the objective of a honeypot is to detect abnormal behavior, such as aggressive flows of traffic. Once the abnormal behavior is detected, it is characterized, for example identifying the flows which are aggressive. The identified traffic is then sent to a dedicated network. This technique is usually called "bait and switch", and allows choosing what is supposed to reach the production network or the honeypots network.

The concept of such a honeypot is based on a gateway that acts both as a firewall and an intrusion detection system (IDS) or an intrusion prevention system (IPS). It filters the flow of traffic arriving to a network, analyses the content of the packets, and then evaluates if the source of a network session is malicious or not by comparing packets to signatures of known attacks. After this identification is carried out, the gateway can mark specific sources as unsafe for a chosen period of time. Thus, packets sent by marked sources will be redirected to the honeypot network rather than to the production network. Instead of dropping the traffic, the traffic coming from source marked unsafe has been retained and can be used for forensics.

For example, in the case of the worm MSBlast [5], if the gateway matches incoming TCP packets with a destination port 135 and has signs of the worm as determined by IDS signatures, it will be able to redirect them to a honeypot while marking the source as contaminated. The honeypot will then manage the future discussions coming from this new, freshly captured worm.

There are some drawbacks with the above technique:

1. Signatures always appear a bit late for new, unknown aggressions.
2. This is not a reliable concept. For e.g., a system can make errors and send some legitimate traffic to the honeypot after a bad interpretation due to a false positive.
3. Network speed may slow down due to the huge work of packet analysis on the gateway.

Role of Honeypots and the Payloads of Worms

In the process to catch the worms and analyze them, honeypots are very useful in dealing with payloads. For e.g., if a worm is infecting a host (which is actually a honeypot), the host (honeypot) can either be a "sacrificial lamb" (a normal host without the very latest updates applied on, sacrificed in expectation of an attack) or just a simulation of services.

Sacrificial Lamb

When using a sacrificial lamb, a classical clean computer can be installed with the required services on the appropriate operating system. This machine can also be part of a system accommodating several operating systems at once, through VMWare [6].

1. To catch a worm, a defenseless machine is left and waits for an infection. Once the worm infects a machine, it is possible to analyze at the worm itself through the new binaries found on the disc and the captured network traffic.
2. When worms are made of only a small amount of code and are not very complex, it is trivial to analyze the network traffic and the binaries caught. But if the worm is sophisticated and involves more complex packets at the network level (cipher), or if it modifies its behavior and signature (similar to viruses with polymorphism), the analysis becomes complicated. Moreover, if the worm quickly destroys infected hosts, the sacrificial lamb could be lost and all posterior and forensic analysis could be compromised.

Virtual hosts and services

By simulating hosts or services, a honeypot can try to dialog with remote incoming worms through the use of its fake services. The daemon called Honeyd [7] is really sharp for such actions.

Case 1

This case presents a configuration allowing one to fool the MSBlast worm, making it think it had contaminated a true Windows host through the RPC DCOM service running over TCP port 135. This simplified configuration was used with success to recover the MSBlast worm over the Internet:

```
create default
set default personality "Windows XP Pro"
add default tcp port 135 open
add default tcp port 4444 "/bin/sh scripts/WormCatcher.sh $ipsrc
$ipdst"
set default tcp action block
set default udp action block
```

And here is the content of the script "WormCatcher.sh" launched by Honeyd for every incoming request coming to TCP port 4444

```
#/bin/sh
# Creation of a directory for every contaminated host
# attacking the honeypot, in order to archive different binaries
mkdir /tmp/$1-$2
# Download of the worm through TFTP in this directory
# (specific behaviour for MSBlast)
cd /tmp/$1-$2/
tftp $1 <<EOF
get msblast.exe
quit
```

Due to this simple technique, binaries of the worm were without even using a real Windows operating system. This technique is therefore rather practical, but it will be difficult to reproduce with worms that use more complex protocols and tools (such as a cipher with encoding keys for each network session). Until recently such types of evil worms have not been widely spread on the Internet.

Role of Honeypots during the Propagation of Worms

Reply to the incoming requests of the worm

When a worm propagates itself, it uses random IP for its future targets. In the list of targets, some IP addresses may be unused, others may not accommodate the remote services targeted by the worm, and some may not answer at all for some or the other reasons. The worm will receive associated network errors and will then try to jump to other targets to spread it as fast as possible.

While replying to a worm's incoming requests toward non-existing hosts or services, a honeypot will be able to steal time from the worm as it attacks its fake targets.

Reply Slowly to the Worm

A honeypot can see incoming requests toward nonexistent IP addresses, services or systems. When a honeypot receives such requests, rather than just answering, it can also reply with specific, crafted packets and slow responses [8] at the network layer [9]. This technique is used with the tool called Labrea [10]. The Unix daemon is able to reply with the unused IP addresses of a network, in order to simulate a TCP session with the attacker. Then it slows down to maximize the duration of the TCP session. The TCP window size is permanently kept to 0, preventing the worm from sending data.

Usually a worm, using classic network calls of the kernel, runs a process on the attacking host, to spread itself elsewhere. By using the classic API, such as sockets, a worm will not be able to understand why the network works slowly, and will be slowed down and blocked on many fake targets. If a worm designer wanted to handle such situation, it would have to design the code for worm more efficiently. Such a worm would probably operate slower (multiple checks would need to be added), become heavier (requiring more extensive code), and with stealth issues (low level access). As an experiment, Labrea had been conceived to fight off the Code Red worm [11].

There are drawbacks to this defense system. For example, a multi-threaded worm, or a time aware worm, may simultaneously attack several targets, without remaining blocked on fake targets.

Honeypots that are configured and specifically "warned" about a worm attack will be able to urgently ask for separation actions. Some other interesting countermeasure actions launched by honeypots could be cutting off a network segment, isolating one or several hosts, closing remote, sensitive services used by the worm, banning chosen network flows by inserting filtering rules on remote devices (routers, firewalls, etc.).

4.2.2 Virus Attacks

A computer virus is a computer program that can copy itself and infect a computer. The term "virus" is also commonly but erroneously used to refer to other types of malware, including but not limited to adware and spyware programs that do not have the reproductive ability. A true virus can spread from one computer to another (in some form of executable code) when its host is taken to the target computer; for instance because a user sent it over a network or the Internet, or carried it on a removable medium such as a floppy disk, CD, DVD, or USB drive. Most conventional anti-virus programs use "signatures" to identify and block viruses. But virus must be first analyzed before sending out the fix. This means that rapidly spreading viruses can cause widespread damage before being stopped. Some researchers have developed artificial "immune systems" that automatically analyze a virus, and therefore a solution can be sent out more rapidly. In practice, however, computer viruses still tend to spread too quickly.

Part of the problem is that countermeasures sent from a central server over the same network where the virus resides will always be risky. The solution is to develop a network of "honeypot" computers, distributed across the Internet and dedicated to the task of combating viruses. To a virus, these machines would seem like ordinary vulnerable computers. But the honeypots would attract a virus, analyze it automatically, and then distribute a countermeasure.

Healing Hubs

Healing hubs are by far the fastest way to counter the virus attacks using honeypots. In healing hubs, the honeypots are linked to one another by a dedicated and secure network. As soon as a honeypot captures a virus, all the others connected to hub will quickly know about the infection immediately. Each honeypot then acts as a hub of healing code which is disseminated to computers connected to it. The countermeasure then spreads out across the broader network. In solutions like healing hub, security measures, such as encryption, are needed to prevent viruses from exploiting the honeypot network.

Case 2

As quoted by Jeff Kephart, a computer scientist at IBM in Hawthorne, New York, US

"They've shown it is possible to use this epidemically spreading immune agent to good advantage, ... the next step would be to look more carefully at the benefits and costs of this approach. I see promise in it."

Simulations show that the larger the network grows, the more efficient this scheme should be. For example, if a network has 50,000 nodes (computers), and just 0.4% of those are honeypots, just 5% of the network will be infected before the immune system halts the virus, assuming the fix works properly. But, a 200-million-node network—with the same proportion of honeypots—should see just 0.001% of machines get infected.

Quarantine Net (QNET) Architecture

Quarantine Net, as the name suggests, quarantines or isolates the computer systems when they are found to be infected with a virus or worm. Figure 4.2 shows the schematic overview of QNET architecture.

QNET architecture is composed of following elements:

Ordinary users: Users of systems that are not quarantined (thus, the big majority of users) are not hindered in their Internet access; their traffic is not impacted by QNET in any way.

Honeypot: The honeypot resides in and listens to the same network as the users. When the honeypot notices a new infection, it triggers the QNET Core, which takes care of isolating the infected system.

Quarantined users: Network traffic from systems that are quarantined is redirected by the router to a QNET server. QNET makes sure that a quarantined system is still able to connect to several "whitelisted" systems by "proxying" the corresponding network traffic.

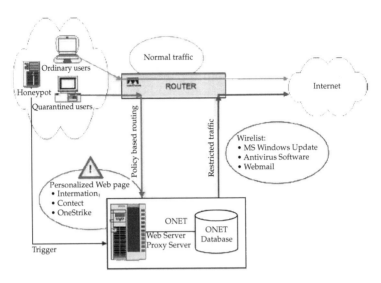

Figure 4.2 Schematic overview of the entire QNET system.

4.3 SPAM AND PHISHING MAILS

As the postal mailboxes are filled with advertising flyers, email inboxes around the world everyday are filled with millions of emails called spam. Spam can be considered as the most annoying cyber-pollution that targets everybody with tons of unsolicited emails. Those emails usually contain advertisements and spammers are paid to spread as many of them as possible.

Though spam itself is generally not considered a real cyber attack, it is difficult to distinguish between virus-contaminated emails, phishing scams and bothersome ads (those containing tricky JavaScript or specific forged HTML used to track them). Moreover, spammers slow the servers receiving legitimate emails and cause availability problems. While spammers earn money by embarrassing people, employees and net surfers lose time by receiving unsolicited emails. Companies lose money through lost productivity, bandwidth charges, purchasing blacklists, etc. Typical solutions against this cyber-plague may be to filter emails received by using content analysis or blacklists, and to fix poorly configured servers.

4.3.1 Spams

Honeypots are very useful to fight the spammers. During the working of the spammers, honeypots can be used for defense in various phases to detect, slow and stop the activities, thus promoting clean Internet.

Working of Spammer

The spam is sent by spammers because the spammers get paid by the activity of cyber mass advertisement. Spammers' work can be divided into different categories:

- Harvest: build a database of targets by finding valid email addresses.
- Stealth and open proxies: work anonymously while sending dreadful emails to their targets.
- Spam and open relays: find and use servers that accept to relay emails anywhere.

Role of Honeypots in Harvesting

First step of spammer is harvesting. Harvesting gives the spammer an updated list of targets. There are many ways to collect thousands of email addresses on Internet. When one sends emails, address will be available to simple, automatic programs that are looking at the headers of every message posted. By saving specific fields (From: Reply-To:), spammers may easily build huge lists of potential targets. Another example of harvesting addresses may be through the use of poorly configured mailing lists that give out the list of their subscribers [11]. A third technique is shown in Figure 4.3. It is based on simple, automatic programs, like crawling Web pages on Internet. For each HTML Web page found, such a program will check for a mailto: link and will follow the Web links.

The harvesting is mainly done through Web pages. The honeypots can be used to fool the spammers during harvesting. The web pages can be loaded with fake e-mail addresses. While spammers browse the Web, if they read web pages with fake email addresses, they will feed their database with invalid targets. Though this does not exactly correspond to a honeypot, it resembles a honeytoken, like adding one spoon of honey on the Web pages.

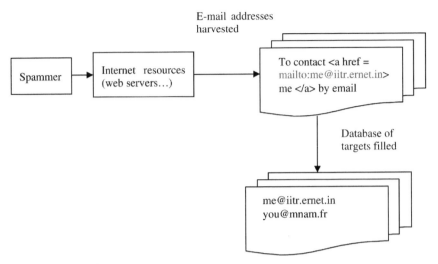

Figure 4.3 E-mail Harvesting.

During automatic harvesting of valid email addresses on the Web, spammers may sometimes be recognized because of the tools they use by checking the *User-Agent* field sent by their browser. Some people have decided to either block a specific *User-Agent* known to be used by spammers, or transparently redirect those Web clients to fake Web pages containing tons of fake email addresses. The trouble is that it is very easy for spammers to change the *User-Agent*. So to defend against spam, it has been decided to create Web links on pages that would be invisible for a human reader (e.g., white characters on a white background) but visible for a spambot following every link read in the HTML source. Such a Web page waiting for Spam bots will dynamically create fake email addresses to fool the spammers.

Though these techniques seem to be interesting, they will only work with naive spambots, ones which are probably not used by skilled spammers. The more sophisticated spammers may use open proxies to crawl the net, and the dynamically created email address will just help with finding such proxies and the spammer will keep his anonymity.

Case 3

One idea could be to create tons of fake addresses. There is a quite good example of a piece of freeware called Wpoison. This CGI script added to your Web site will generate fake email addresses looking like real ones. Another technique could be to create a fake address containing specifically chosen information. The day this email address is used as a target of spam, the owner will be able to determine the IP used by the spammer.

```
<?
// PHP example taken from the frenchhoneynet Web site
// replace by your domain, add recipients filtering on your MTA
(mimedefang...)
echo '<a href="mailto:'.$REMOTE_ADDR.'_'.date('y-m-j').'-spamming@frenchhoneynet.org"
title="There is no spoon">For stupid spambots';
?>
```

This script will dynamically generate a mailto: link, containing a fake email address with the IP of the current Web client and the date. For example:

```
<a href=mailto:80.13.aa.bb_03-11-17-spamming@frenchhoneynet.org>...
```

If the Web client is a spambot, it will add 80.13.aa.bb_03-11-17-spamming@frenchhoneynet.org in the database of potential targets. Now suppose that a spammer uses this database. He will probably send an email to this virtual address.

Then the mail server administrator can filter incoming emails by looking at the recipients (on your MTA or eventually on your MUA [Mail User Agent]). If one receives an email destined to 80.13.aa.bb_03-11-17-spamming@frenchhoneynet.org, then one is sure that 80.13.aa.bb is the IP address that was used on November 17, 2003. And more than that, one knows that this address was a spam harvesting source.

```
# Example of a simple recipient filtering with Mimedefang http://
www.mimedefang.org/]
# Will filter incoming email containing a recipient address in the form
# of those created by the latter PHP example.
sub filter_recipient {
        my ($recipient, $sender, $ip, $hostname, $first, $helo) = @_;
        if($recipient =~ /^<.*-spamming@frenchhoneynet\.org>?$/i)
        { return ("REJECT", "Spamming activity"); }
        return ("CONTINUE", "ok");

}
```

Honeypots and Open Proxies

Spammers may either directly connect to a remote mail server, or bounce through open proxies. For example, the role of a Web proxy is to do the job of a Web client for someone else. When a Web client connects to a proxy, he asks for a Web page somewhere on Internet. The proxy will then grab this Web page by itself, and will return the obtained data to the client. In the logs of the remote Web server, usually one can only see the IP address of the proxy who did the Web requests.

An open proxy is a proxy service opened to the world for almost any kind of request, allowing anybody to remain anonymous while crawling the net. Such proxies are used a lot by the attackers. Open proxies are also useful for many spammers, because they will be able to stay anonymous while sending their unwanted emails.

Using a proxy server is quite efficient for a spammer to have anonymity. As proxy owners may have logs, spammers may fear that their IP address could be recorded (remote proxy log). Usually, badly configured proxies don't have logs. Their fear of logs is why sometimes they use chains of proxies to increase complexity, as shown in Figure 4.4. They connect to an open proxy server (TCP Session), then ask it to connect to another known open proxy server (CONNECT a.b.c.d:3128), etc.

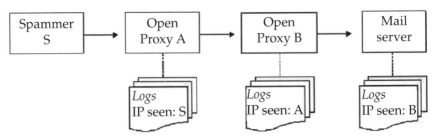

Figure 4.4 Open relays and spammers.

The longer the chain, the stealthier they become, but they will lose time as multiple bounces will result in multiple delays added.

One of the main paths used by spammers to reach mail servers is going through open proxies that accept and freely transmit requests.

These open proxies play the role of screeners for the spammers that hide beyond them. A fake open proxy is set up in a Honeypot as shown in Figure 4.5. By connecting to the answering TCP ports, sending a few packets may help to understand if the proxy is open or not. Honeypots are set up that will answer positively to incoming requests and fool some spammers as shown in Figure 4.6.

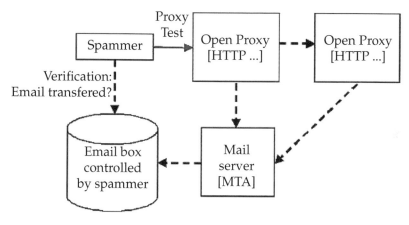

Figure 4.5 Phase one—spammer checks the open proxy.

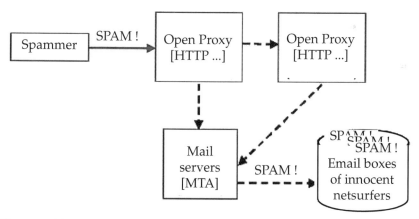

Figure 4.6 Phase one—spam.

Case 4

Here is an example of a TCP session recorded by snort showing a remote proxy check probably launched by Earthlink. The client connects to the proxy on TCP port 8080, and doesn't ask for a Web page but instead for a TCP session initialized with a remote SMTP server (207.69.200.120) owing to the HTTP CONNECT function. The rest of this TCP session is SMTP, directly sent to the SMTP server (HELO, MAIL FROM, RCPT TO, DATA, QUIT).

$ cat /var/log/snort/192.168.1.66/SESSION\:8080-4072

CONNECT 207.69.200.120:25 HTTP/1.0

HELO [217.128.a.b]

MAIL FROM:<openrelay@abuse.earthlink.net>

RCPT TO:<spaminator@abuse.earthlink.net>

DATA

Message-ID: <36af800461754252ab1107386a9cd8eb@openrelay@abuse.earthlink.net>

To: <spaminator@abuse.earthlink.net>

Subject: Open HTTP CONNECT Proxy

X-Mailer: Proxycheck v0.45

This is a test of third-party relay by open proxy.

These tests are conducted by the EarthLink Abuse Department.

EarthLink, by policy, blocks such systems as they are discovered.

Proxycheck-Type: http

Proxycheck-Address: 217.128.a.b

36af800461754252ab1107386a9cd8eb

Proxycheck-Port: 8080

Proxycheck-Protocol: HTTP CONNECT

This test was performed with the proxycheck program. For further information see http://www.corpit.ru/mjt/proxycheck.html/.

QUIT

Honeypots and Open Relays

An open relay (which is sometimes called an *insecure relay* or a *third-party relay*) is a Mail Transfer Agent (MTA) that accepts third-party relays of e-mail messages even though they are not destined for its domain. As they forward emails that are neither to nor from a local user, open relays are used by spammers to route large volumes of unsolicited emails. Figure 4.7 shows use of open proxies and relays by spammers.

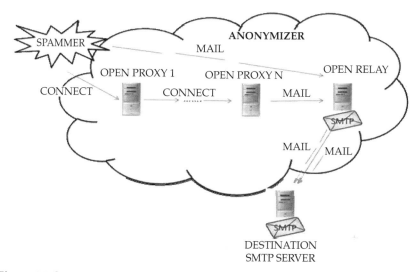

Figure 4.7 Open proxies, open relays and spammers.

Such a poorly configured MTA lends its system and network resources to the remote abuser who is getting paid to send out spam. Usually, an organization that unwittingly relays spam may be blacklisted on international lists (RBL, etc.). That would annoy internal users because they couldn't use their own email properly. A big ISP sadly blacklisted would probably lose clients and money.

One can create a fake mail server (honeypot) and transform an unused sendmail daemon to fool spammers. This can be easily done by asking sendmail to accept relaying and to queue every email without ever sending one email out. This configuration offers a service that looks like a real open relay.

Architecture of HoneySpam

Figure 4.8 shows the architecture of HoneySpam. The framework is placed into the DMZ area of an enterprise network. HoneySpam includes following emulated components: *fake web servers* (to allow e-mail harvesting), *fake open proxies, fake open relays* (to provide spammers with services that can be used to their advantage), *destination SMTP servers* (to increase spammers traceability). Service emulation is obtained through honeypots.

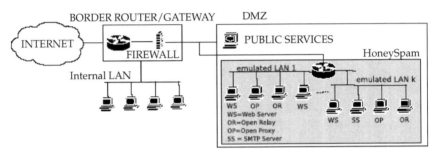

Figure 4.8 Architecture of HoneySpam.

Fake Web server: The pages of HoneySpam Web servers provide "special" email addresses that, if used by the spammer, help trace him back. Furthermore, each action is logged in order to understand who is accessing the Web servers. The *fake Web servers* generate dynamic pages containing many links and e-mail addresses which cannot be identified as fake by common spammer tools. The number of links, as well as that of the addresses is randomly chosen in a given range.

Fake open proxies and relays: In order to destroy the anonymity of the spammer, virtual open proxies and open relays are provided to intercept illegal traffic operated by spammers. Connections to open proxy and open relay servers are logged, to understand where the unwanted traffic is coming from. Furthermore, the traffic is blocked, thus reducing the number of unwanted e-mails reaching target users.

Fake destination SMTP server: The destination SMTP server is specific to the domains protected by HoneySpam. If the spammer has harvested e-mail addresses from Web servers present in HoneySpam, these addresses will correspond to mailboxes in HoneySpam's destination SMTP servers. The idea is to redirect all the traffic coming to any of the special address to a single mailbox, in which all the messages are logged and stored for analysis and backtracking purposes. The server is configured not to send any message, if requested.

4.3.2 Phishing

The term phishing (password harvesting fishing) describes the fraudulent acquisition, through deception, of sensitive personal information such as passwords and credit card details by masquerading as someone trustworthy with a real need for such information. Victims often suffer significant financial losses or have their entire identity stolen, usually for criminal purposes. Figure 4.9 shows steps in a phishing attack.

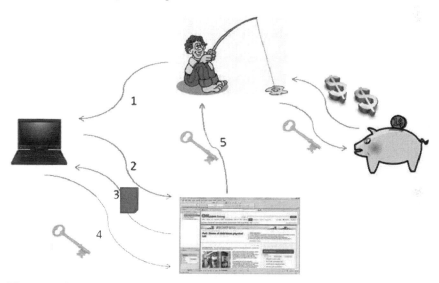

Figure 4.9 Steps in a Phishing Attack.

All phishing attacks fit into the same general information flow. At each step in the flow, different countermeasures can be applied to stop phishing. The steps are:

1. The phisher prepares for the attack. Step 1 countermeasures include monitoring malicious activity to detect a phishing attack before it begins.

2. A malicious payload arrives through some propagation vector. Step 2 countermeasures involve preventing a phishing message or security exploit from arriving.

3. The user takes an action that makes him vulnerable to an information compromise. Step 3 countermeasures are detecting tactics and rendering phishing messages less deceptive.

4. The user is prompted for confidential information by a remote web site. Step 4 countermeasures are focused on preventing phishing content from reaching the user.

5. The user compromises confidential information. Step 5 countermeasures concentrate on preventing information from being compromised.

6. The confidential information is transmitted from a phishing server to the phisher. Step 6 countermeasures involve tracking information transmittal.

7. The confidential information is used to impersonate the user. Step 7 countermeasures center on rendering the information useless to a phisher.

8. The phisher engages in fraud using the compromised information. Step 8 countermeasures focus on preventing the phisher from receiving money.

Honeynet technology captures in detail the typical life cycle of a phishing attack. The information is captured using high interaction research honeypots. In each incident phishers attacked and compromised the honeypot systems, but after the initial compromise their actions differed and number of techniques for staging phishing attacks were observed:

1. Setting up phishing web sites targeting well known online brands.

2. Sending spam emails advertising phishing web sites.

3. Installing redirection services to deliver web traffic to existing phishing web sites.

4. Propagation of spam and phishing messages via botnets.

Usage of honeypots in network demonstrate that phishing attacks can occur very rapidly, with only limited elapsed time between the initial system intrusion and a phishing web site going online with supporting spam messages to advertise the web site, and that this speed can make such attacks hard to track and prevent.

4.4 DISTRIBUTED DENIAL OF SERVICE ATTACKS

A denial of service (DoS) attack is commonly characterized as an incident in which a user or organization is deprived of the services of a resource they would normally expect to have. Typically, the loss of service is the inability of a particular network service, such as e-mail, to be available or the temporary loss of all network connectivity and services.

Distributed DoS attack is used in order to magnify the effect on the victim. It uses many computers to launch a coordinated DoS attack against one or more targets. Using client/server technology, the perpetrator is able to multiply the effectiveness of the Denial of Service significantly by harnessing the resources of multiple computers which serve as attack platforms. Typically a DDoS master program is installed on one computer. The master program, at a designated time, then communicates to any number of "agent" programs, installed on computers anywhere on the Internet. The agents, when they receive the command, initiate the attack. Using client/server technology, the master program can initiate hundreds or even thousands of agent programs within seconds. Figure 4.10 shows the overview of DDoS attack scenario.

A DoS attack's main characteristics are that an attacker attempts to prevent one or more legitimate users of a service from the use of the required resources. Therefore, he attempts (1) to inhibit legitimate network traffic by flooding the network with useless traffic, (2) to deny access to a service by disrupting connections between two parties, (3) to block the access of a particular individual to a service, or (4) to disrupt the specific system or service itself.

DDoS attacks follow the same path, but they become more effective and difficult to prevent because of the intermediary systems that add a many-to-one dimension to the whole attack. A popular attack misuses the standard three-way handshake of the Transport Control Protocol (TCP). This handshake requires an exchange of three messages between client and server before the service can be used: (1) The client indicates that it wants to start a connection

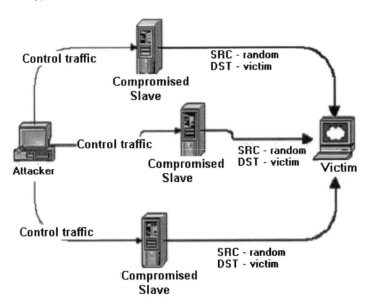

Figure 4.10 Overview of DDoS attack scenario.

with the server by sending a synchronize (SYN) request. (2) The server replies with a message indicating its readiness: a SYN/ACK (Acknowledgment) reply. (3) The connection can be used after the final ACK of the client. The so called SYN flood attacks the server—the victim in this case—by sending a large amount of SYN requests without fulfilling the third step of the handshake. Typically, the client's address is also spoofed. Because today's TCP/IP implementations only handle a limited number of connections, the server will discard new connections as long as its backlog queue is full with semi open connections.

Role of Honeypots in Defending Against DDoS

The advantages of using honeypots for DDoS are two-fold: First one can defend operational network with a high probability against known DDoS and against new, future variants. Second, one can trap the attacker so that recording of the compromise can help in a legal action against the attacker. Honeypot lures the attacker to believe that he successfully compromised a slave for his needs. In reality, the honeypot learns the tools, tactics, and motives of the blackhat.

The lessons learned are then implemented in the rest of the network as defensive mechanisms. The honeypot provides organization information on their own security risks and vulnerabilities. It should consist of similar systems and applications than the one used by the organization for its productive environment so to give the attacker a real world feeling and to be able to implement the learnt lessons in the productive environment. Figure 4.11 illustrates the implementation in the organization.

Until now, the use of Honeypots has been focused as a Detection System. However, Honeypots have also emerged as a Deception System, differentiating them from traditional security systems, and protecting the real, effective information from attackers. Here instead of using Honeypot to lure the attacker, they are being used to threaten the attackers of some trap. The purpose of deploying honeypot is not to be compromised by some attacker but to deceive him of some trap and protect the real information. Attackers are usually intentionally presented with host(s) on the network that has one or more vulnerabilities. This creates suspicion about the presence of Honeypot and protects the real asset.

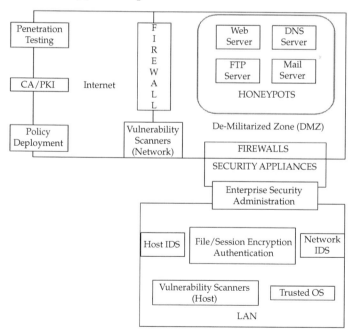

Figure 4.11 Implementation in the organization.

The organization shown in Figure 4.11 is composed of both classical and recent security elements, thus providing a well designed security infrastructure. Demilitarized zone(DMZ) consists of web server, FTP and DNS servers for access from outside the organization. The LAN is protected by firewall and other security appliances from the outside world. Internally, also, the file transmission is encrypted, the clients run trusted operating system and authentication is provided for controlled usage. Apart from this, Intrusion detection system (Network and Host) and vulnerability scanners are installed in the LAN to detect any suspicious activity. Standards security techniques are employed at the web and mail server. In this infrastructure, a honeypot attracts the distributed denial of service attacks. It simulates the whole network as shown in Figure 4.12, to lure the attacker into employing it as a compromised slave and then handles its packets.

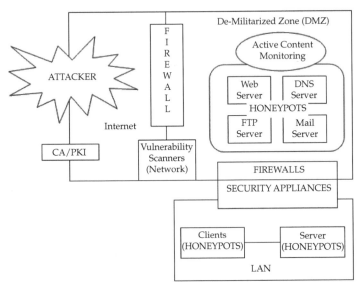

Figure 4.12 View of the attacker.

Every system in the organization might be a honeypot. For example, if the attacker's packets to the web server of the corporation are detected, the packets go to the honeypot for processing. The reply the attacker gets should be indistinguishable from a real reply of the web server. Three major problems must be solved to successfully project this illusion to the attacker:

1. The attack must be detectable.
2. The attack packets must be actively directed to the honeypot.
3. The honeypot must be able to simulate the organization's network infrastructure, at least the parts known to the attacker.

The first issue is linked to the solution of the second problem: both should ideally be implemented by a transparent packet forwarder at the border of the corporation's DMZ.

Its functionality is to look at each packet and decide if the packet belongs to a DDoS attack. If the test is negative, the packet should go to the given destination inside the DMZ or the LAN. In all other cases the packet forwarder should determine which part of the honeypot system should produce the request. A possible setup of the honeypot could be that each Internet service of the corporation is replicated in one system of the honeypot. The detection itself is done by efficiently matching signatures of DDoS packets. Currently, one employs similar signatures as the DDoS signatures to be able to detect known attacks with a large probability. Finally, the third problem can be solved by employing a variant of the approach. Then, it should also be easier to simulate realistic confirmation messages to the attacker as depicted in Figure 4.13.

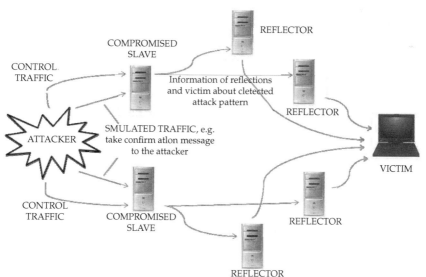

Figure 4.13 Tracing attacker.

The depicted warning system of the honeypots to the reflectors and the victim enables to play down eventual probes of the attacker to verify the success of the DDoS attack at these points.

4.5 SUMMARY

Honeypots can be used for prevention, detection and reaction to attacks. Honeypots are a valuable additional means to fight against virus and worms. Honeypots are an effective mean to detect spammers, slow spammers and even block the spammers. Honeypots have also proven to be successful against phishing attacks. The real value of honeypots lie in detecting and mitigating distributed denial of service attacks, which are a real and most costly threat to the security of the Internet today.

EXERCISES

Short Answer type Questions:

1. List the generic phases of the attack.
2. What is the difference between a worm and a virus?
3. How can honeypots be useful to fight against worms?
4. List the components of a honeyspam architecture.
5. What is the difference between DoS and DDoS?

Long Answer type Questions:

1. Discuss the countermeasures that can be taken in each phase of an attack. What is the role and value of honeypots in each attack phase?
2. Explain the lifecycle of a worm.
3. According to you, which stage of worm can best be countered by honeypots and why?
4. Explain the various components and working of quarantine net architecture.
5. What do you think are the basic problems in the QNET architecture and propose a solution to modify the Q-Net architecture?

6. Explain how honeypots can be used to protect against spams and phishing.

7. Discuss the components of HoneySpam architecture.

8. Think about the various situations of DDoS in network and outline network diagrams to defend against DDoS using honeypot in each case. You should cover situations that require production honeypots in DMZ, inside firewall, outside firewall of combination thereof.

REFERENCES

[1] L. Spitzner, *Honeypots: tracking hackers*: Addison-Wesley Professional, 2003.

[2] C. Stoll, *The cuckoo's egg*: Pocket Books, 1989.

[3] R. Caddington (Eeye), "Decoding and Understanding Internet Worms," 2001. http://www.blackhat.com/presentations/bh-europe-01/dale-coddington/bh-europe-01-coddington.ppt.

[4] G.A. Edward, *Fundamentals of computer security technology*: Prentice-Hall, Inc., 1994.

[5] L. OUDOT, "Security advisory for the MSBlast worm," 2003. http://www.microsoft.com/security/security_bulletins/ms03-026.asp.

[6] "VMWARE," http://www.vmware.com.

[7] N. Provos, "Honeyd—A Virtual Honeypot Daemon," in *Proceedings of the 10th DFN-CERT Workshop*, Hamburg, Germany, Feb. 2003.

[8] T. Bautts, "Slowing down Internet worms with tarpits," 2003. http://www.securityfocus.com/infocus/1723.

[9] Z. Chen, L. Gao, and K. Kwiat, "Modeling the spread of active worms," in *Proceedings of Twenty-Second Annual Joint Conference of the IEEE Computer and Communication (INFOCOM 2003), IEEE Societies.* vol. 3, 2003, pp. 1890–1900.

[10] T. Liston, "Welcome to my tarpit: The tactical and strategic use of LaBrea," *Dshield. org White paper*, http://labrea.sourceforge.net/labrea-info.html, 2001.

[11] CAIDA, "Caida Analysis of Code-Red". http://www.caida.org/research/security/code-red/.

5

Static Honeypots

This chapter describes the simplest form of honeypot i.e., static honeypot. They can be used as detection systems or as deception mechanisms. As a detection mechanism, discusses the requirements of the honeypot framework. It provides an overview of different layers of the Japonica framework and discusses the components of each layer. Finally, the chapter outlines how honeypots can be used for increasing deception and introduce Deception Tool Kit (DTK) and Deception Wall (D Wall).

Static Honeypots systems are the one in which the number and location of Honeypots are fixed. There can be two ways in which Honeypot can be used as a network security tool—as a detection system and as a deception mechanism. As a detection system, honeypot is deliberately made vulnerable with fake services to lure the attacker towards it. Intruders are expected to compromise the discovered honeypot. The traces left by the intruder are then used by the system administrators to investigate and learn about their tools and techniques in detail. It is used to complement the functionality of a network IDS [1]. The honeypot is coordinated proactively with the firewall and the IDS, to achieve early response to network security incidents. As a deception mechanism, Honeypot provides the attacker with what he needs, and influence opponent choices in one's own favor to consume the attacker's resources in the process.

5.1 HONEYPOT AS DETECTION SYSTEMS

This section discusses the objectives and requirements of Honeypot Framework as a Detection System, the Japonica framework [2], that achieves early response through the dynamic orchestration of prevention, detection, and response mechanisms. The word Japonica is derived from the scientific name of the Japanese honeybee—*Apis cerana japonica*.

5.1.1 Japonica: Objectives and Requirements

The basic objectives of a Honeypot frameworks in general, and Japonica in specific are:

Early and complete response to unknown attacks: The overall goal of Honeypot framework is to actively detect and respond to unknown attacks. The detection should be as early as possible, preferably before the victim resources are harmed. It also aims to prevent future attacks of similar kind. The response needs to be *complete* so that attacks would be contained properly. The attack should be addressed by preventative, detection, and response measures.

Accurate detection of attacks: Current network IDSs still have problems distinguishing between legitimate traffic and malicious traffic, especially if they demonstrate similar patterns. The framework should aim to minimize incidents where legitimate traffic is mistaken to be malicious traffic. Therefore, it is required to minimize false positives as much as possible.

Scalability: Due to the diversity of security technologies, scalability is a crucial objective. Any intrusion detection system or firewall can be deployed along with honeypot. With the scalability of the Japonica framework, any intrusion detection system or firewall can be deployed within Japonica. To achieve this objective, a uniform message exchange format between different components has to be defined. Such a format would enable the different components to communicate with one another.

To achieve the above mentioned objectives, a honeypot framework should meet the following requirements:

Real-time coordination of prevention, detection, and response mechanisms: The first objective of the framework is focused on early and complete response to unknown attacks. In order to achieve this, all three distinct aspects of security i.e., prevention, detection, and response, need to be considered. More specifically, these three mechanisms need to be coordinated in real time so that attacks can be responded to early. The framework would require input from distinct sources from the detection mechanisms, and invoke the prevention and response mechanisms accordingly.

Generic mechanism to address a wide range of attacks: To focus unknown attacks, the framework should not just accommodate for attacks that conform to a specific pattern but should also detect malicious patterns.

Dynamic and risk-aware feedback based approach to detect and respond to attacks: The framework needs to be risk-aware. Risk itself is not a static concept, but a dynamic one, especially in complex real world environments with many variables to consider. The dynamism of such environments needs to be a critical factor in the framework. To achieve dynamic risk assessment, one important technique that can be used is to make it feedback-based.

Use of a mechanism that does not exhibit false positives: The second objective states that false positives should be minimized. Hence, the framework would need to use a security mechanism that does not give a huge number of false positives.

Uniform message exchange format: A uniform message exchange format needs to be designed. It is to enable different components, including new components, to communicate with one another within the framework.

5.1.2 Framework and Components

In Japonica framework [2], there are basically *seven* components organized in four layers. The various components are (i) Compromise Detection Component (CDC), (ii) Network Traffic Sensor (NTS), (iii) Compromise Analyzer (CA), (iv) Threat Evaluation Component

(TEC), (v) Response Coordinator (RC), (vi) Prevention Engine (PE), and (vii) Detection Engine (DE). These components are organized as a layered stack, for clarity and modularity, which would in turn facilitate implementation. Figure 5.1 shows the various components of the framework organized into four layers.

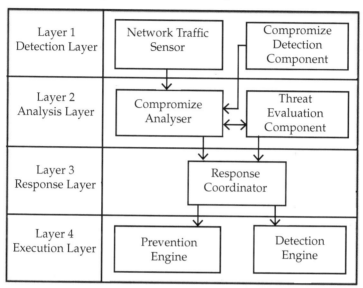

Figure 5.1 Honeypot Framework.

Detection Layer (Layer 1): The main purpose of honeypot is to collect information related to intrusions on the host. The first and the most important layer, the Detection Layer, performs this job of collecting all kinds of information. This Layer has Compromise Detection Components (CDC) and the Network Traffic Sensor (NTS). CDC Module runs on the host and detects whether the host is compromised or not. If any such activity occurs it records about the specific features of the compromise. Different CDCs may have different functions to detect various kinds of intrusions. Some of the functions of CDCs are:

1. Detecting whether there is any rootkit downloaded to the host.
2. Detecting any unusual port that is listening on the host.
3. Detecting any accounts that are created without authorization on the host.
4. Detecting any critical files that have been replaced by an attacker.

A CDC by itself can perform either a simple lightweight function, or a complex heavyweight function. However, when one runs many CDCs with different functions together on a single host, the situation becomes increasingly complex. Therefore, organizations generally adjust their additional CDCs according to the requirements of specific environments. For example, in the environment of a financial institution, a CDC can be designed to detect stolen credit card numbers. The Network Traffic Sensor component keeps the records of suspicious network traffic, including source port, destination port, source IP, destination IP, protocol type and packet size.

Analysis Layer (Layer 2): The information collected in Layer 1 is sent to the second layer, the Analysis Layer, to analyze and determine whether the input is a malicious activity or not based on the threat level and the threshold value. The threat level is a quantifiable measure of the risk associated with a set of one or more CDCs. Each unique set of possible CDCs will have its own threat level. If system has n CDCs, the number of possible threat levels would be 0 to $2^n - 1$. The more suspicious the entity behaves, the higher the threat level will be. The threat level determines the risk associated with an entity. Risk is calculated as Probability of Occurrence of threat Level. For example, if system has 2 CDC functions: detection of rootkit download and detection of any unauthorized account creation, then we have 4 possibilities:

1. Threat Level 0: No rootkit download, and No unauthorized account creation.
2. Threat Level 1: A rootkit is downloaded in the host.
3. Threat Level 2: An unauthorized accounts is created.
4. Threat Level 3: A rootkit is downloaded and an unauthorized account is created.

Assume that probability of occurrence of threat level 0 is .8, probability of occurrence of threat level 1 is .1, probability of occurrence of threat level 2 is .075 and probability of occurrence of threat level 3 is .025. So, in the above example, Threat Level 2 is more risky than Threat Level 3. To capture the dynamism of the environment, the threat levels can be made dynamic—They will change dynamically based on the risk that is perceived in the environment.

The **threshold** represents the tolerance towards suspicious events. Unlike the threat level which is dynamic, the threshold is statically defined. Once the threat level exceeds the threshold, it is considered as a high risk intrusion. High risk Intrusions are then sent to Layer 3 to generate rules out of them.

Response Layer (Layer 3): Layer 3 is the Response Layer. It consists of only the Response Coordinator. Layer 3 of Japonica is an intermediary layer, and its main purpose is to generate response rules for Layer 4. According to the analysis results and information input from Layer 2, the Response Coordinator generates corresponding prevention rules and detection rules and sends them to the Prevention Engine and Detection Engine on hosts under attack and also notifies system administrators.

Execution Layer (Layer 4): The Prevention Engine and Detection Engine are located in Layer 4, the Execution Layer. The main feature of the Prevention Engine is to filter out and block malicious threats, and the Detection Engine is responsible for detecting any ongoing attacks based on the detection rules generated by the Response Coordinator. The Detection Engine of layer 4 is signature based, i.e., it detects the attacks in the network based on static pattern matching functionality. This detection process is different from detection in Layer 1, which is mainly host-based and is dynamic in nature. The dynamic rule adjustments are processed in Layer 2, and the unknown attack detection functionality is carried out in Layer 1.

Example, explaining the overall Japonica Model: When a host is under attack from the CodeRed II worm, a file named "root.exe" will be added to the host. At this point, one of the CDCs on that host that is tracking file integrity will send out a message to the Compromise Analyzer. The threat level of the downloaded file will be set, say 5, according to the threat level policy defined by the administrator. After the host is infected, CodeRed II will attempt to connect to other hosts using generated random IP addresses. The CDC will track all outgoing network traffic, and then send a message to Compromise Analyzer. At the moment, the current threat level will increase; say from 5 to 12 according to threat level policy. After this stage, the Compromise Analyzer would inquire the Threat Evaluation Component to find out the current threat level policy. Say the administrator defined the threshold to be 10. Since the threshold has been exceeded, the Compromise Analyzer sends the file information and outgoing traffic information to the next layer, the Response Layer. The Response Coordinator in the Response Layer, generates the detection rules according to the file information and outgoing traffic information from the Compromise Analyzer. It sends the detection rules to the Detection Engine. The Response Coordinator also generates prevention rules and sends them to the Prevention Engine. After the Detection Engine receives detection rules from the Response Coordinator, it will reset the engine to apply the new detection rules. The Prevention Engine will also apply new prevention rules. Finally, the Prevention Engine blocks any traffic with the same signature thus stopping the spread of CodeRed II.

5.2 HONEYPOT AS DECEPTION SYSTEMS

Until now, the use of Honeypots has been focused as a Detection System. However, Honeypots have also emerged as a Deception System, differentiating them from traditional security systems, and protecting the real, effective information from attackers. Here instead of using Honeypot to lure the attacker, they are being used to threaten the attackers of some trap. The purpose of deploying honeypot is not to be compromised by some attacker but to deceive him of some trap and protect the real information. Attackers are usually intentionally presented with host(s) on the network that has one or more vulnerabilities. This creates suspicion about the presence of Honeypot and protects the real asset.

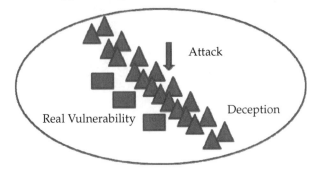

Figure 5.2 The Structure of Defensive Network Deception.

Figure 5.2 shows the basic structure of deceptive defense. There are some real vulnerabilities in the system and some fake vulnerabilities added in the system to deceive the attacker. When attacker comes, he sees so many vulnerabilities. These deceptive vulnerabilities protect the system as:

- Deception increases the attacker's workload because they can't easily tell which of their attack attempts would work and which would fail.
- Deception allows defenders to track attacker attempts at entry and respond before attackers come across vulnerability the defenders are susceptible to.
- Deception exhausts attacker resources.
- Deception increases the sophistication required for attack.
- Deception increases attacker uncertainty.

So, the real vulnerabilities are protected from being exploited.

The Deception Technique

The Deception Mechanism is to:

- Detect attacker's intelligence, motives and attack soon enough.
- Influence attacker's choices in defender's favor.
- Consume attacker's resources in the process.

The attacker, with some initial set of beliefs about the defender's system, tries to get success in his attack by reducing the uncertainty about what is present in the defender's systems using his intelligence. According to Shannon [3], the uncertainty depends on the information content. This is Shannon's notion of increasing information content directed toward the defender's systems. The ideal deceptive defense in such case, allows the attacker's believe to proceed in such a manner that the attacker's intelligence effort appears to meet his expectations. In other words, the attacker moves toward increased certainty at an appropriate rate, but the content the attacker achieves in defender's system is to the defender's advantage. Over the long run, the successful defense will induce the attacker to believe that the technical attacks were successful and that some other circumstance was the cause of the ultimate failure of the overall strategy.

Case 1

A classic example of a successful defense of this sort is the deception program carried out in World War II prior to the Normandy landings. In one case, British intelligence created a set of fictions surrounding the landings that fooled Hitler even after the landings took place. It was several days before he figured out that the real landings were not a feint to cover up other landings, and by that time it was too late.

While deploying honeypots as Deceptive Systems [4], the goals are to:

1. Decrease the probability that an attacker finds the real vulnerability. This can be done by increasing the total size of the space to be searched by the attacker, i.e., to increase the vulnerabilities by adding some fake ones.

2. Increase the probability of finding the fake vulnerability. This is achieved by having a large number of deceptions in the space relative to the number of real vulnerabilities. The higher the ratio of deceptions to vulnerabilities, the more likely this is.

3. Make the time to defeat a deception infinite. This is done by making the deceptions extremely realistic and assuring that defeating a deception system does not provide additional undesired intelligence or paths to successful attack.

4. Make the time to detect a vulnerability once a deception is encountered from a given attack location infinite. The notion is that detecting intelligence attempts against deceptions should be very easy and that when they are encountered, further intelligence from the same source should redirect all intelligence attempts against vulnerabilities toward deceptions. The challenge with this strategy is that inconsistencies will then exist between undetected and detected intelligence efforts that might tend to reveal the real vulnerabilities.

5. Make the time to detect an attack against a deception very small. It is reasonable to assume that a deception can be designed to rapidly detect that it is being examined by an attacker. In

practice, technical probes have been easy to detect, however, there are certainly cases of passive intelligence probes where this is not the case.

6. Make the time to react to an attack against a deception very small. Assuming we can detect an intelligence probe or an attack and that the response can be reasonably pre-programmed, rapid reaction time is often feasible, especially against remote technical intelligence efforts.

Deception Tool Kit (DTK) Model

The use of Honeypot as Deception System has emerged in form of Deception Tool Kit (DTK) Model [5].

Early honeypot systems were based on the idea of placing a small number of attractive targets in locations where they are likely to be found, and drawing attackers into them. Figure 5.3 shows the early Honeypot placed in the maze of real systems.

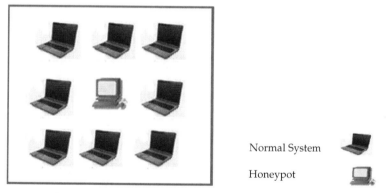

Normal System

Honeypot

Figure 5.3 Early Honey Pot Systems.

The challenge for attackers in these systems is to find a way to detect the honeypot and not to concentrate his efforts against them. This challenge is relatively easy to meet and they could easily adapt to these defense mechanisms. Example, any typical Internet-based Web site defacer can look in advertised locations for high profile systems to attack. It is also effective as a system to 'switch' an attacker to honeypot, once an attack attempt has been detected. So, these are mild ways to consume attackers' resources and time.

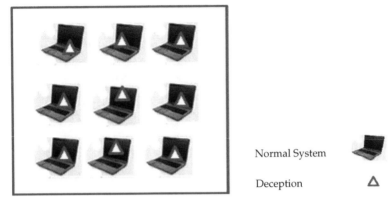

Figure 5.4 Deception spread among normal systems.

Figure 5.4 shows the model of deception spread among normal systems. The fake vulnerabilities are distributed all over the network, among the normal systems, deceiving the whole network as some Honeypot. Deceptions are spread among the normal systems in a network in such a way that unused services on those systems are consumed with deceptions. This has two effects. One effect is that it spreads the deceptions over a larger portion of the IP/port address space, a similar effect to Shannon's 'diffusion' used in cryptographic systems. The other effect is that, this increases the percentage of deceptions in the environment, thus increasing the likelihood of an attacker encountering a deception rather than vulnerability.

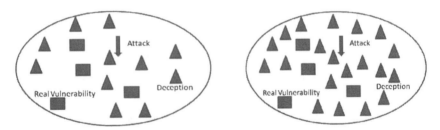

Figure 5.5 Increased Deceptions.

Figure 5.5 shows the model with increased vulnerabilities, thus having an improvement over previous model.

In addition to the above advantages, there are some limitations in the DTK model of deception. Less Deception in DTK had no affect on the overall search space for the attacker's effort. If the deception is of relatively poor quality, it increases the scarcity of real vulnerabilities in the search space only minimally. DTK is only really effective against probes that have still not got way into the network. If an attacker has gotten part of the way into a network and is willing and able to engage in observation rather than active probing, the real services will rapidly become apparent. For this reason, while DTK is effective against more of the current threats in the current environment than conventional honeypots (Detection Systems), it is unlikely to be effective at influencing opponent choices where that opponent has a more advanced intelligence capability.

Therefore it is important to

1. Increase the Size of search space.
2. To enhance the quality of Deception.

Increasing the Size of the Search Space and the Sparsity of Real Vulnerabilities

One way to ease the situation for the defender is to increase the intelligence workload of the attackers by increasing the size of the search space. This can be done in fairly trivial ways, but the results will also be fairly trivial to defeat by attackers of skilled intelligence groups. Still, even a small improvement 'raises the bar', i.e., makes the goals of the attackers more difficult to achieve.

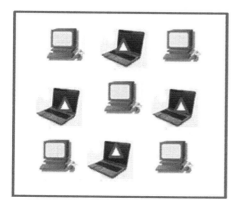

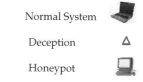

Figure 5.6 Multiple Deceptions in a single box.

Most modern computer systems have multi-homed capabilities. Many IP addresses can be associated with a single Ethernet card, each IP address optionally having its own MAC address as well. This technique can be applied for deception by filling a large address space. The cost of this technique is very low. One can place more than 4,000 IP addresses on a single PC running Linux, which means that with 16 $500 computer systems and about $500 worth of connecting cables and Ethernet hubs (for a total of only $8500 worth of hardware) one can do a deception that covers all of the IP addresses in a class B IP network (a.b.*.*).

To increase the workload of attacker's effort in determining which of these systems are legitimate and which are not, services could be placed on all of these IP addresses. While the deceptions are relatively easy to spot, they are highly effective in causing the attacker's workload to go up, in increasing the time to attack, and decreasing the probability of intelligence probes going undetected.

When a probe encounters a false service, with proper access to outside routers, system can redirect all traffic into deceptions so that subsequent remote access is deception from that point forward. The conditions under which switchovers occur and what services are switched over in what conditions are held confidential. Maintaining confidentiality is important because any single known valid service that could be easily differentiated from a deception could be used to test whether the intelligence probe had been detected and responded to. The sophisticated intelligence effort would then switch to another source location and continue the search.

One can also determine a level at which the IP address space is exhausted before the probability of a probe gets high enough to be of concern. This is done by the following process:

Set RS = the number of real services. (example 1,000)

Set DS = the number of deception services. (example 1,000,000)

Set RRD = RS/DS—the ration of real services to deceptions. (example 1/1,000 = 0.1%)

The initial probability of encountering a real service is RRD (0.1%). For each failed probe that is detected by the attacker, assume a new IP address must be used to continue probing. For each failed probe, only the specific service on the specific IP address can be eliminated from probing. This on the second try, the new value of RRD is given by RS/(DS-1). After 100,000 probes, the probability reaches RS/(DS-100,000) or (1,000/900,000) or 0.11%. Depending on various conditions, different assumptions can be made and the computation becomes more complex, but the notional result is that the resource requirement to gain meaningful intelligence via random probes is extremely high.

These are called intelligence probes because the opponent is taken to be intelligent. A more intelligent approach might be for an attacker to assume that they will first identify machines with seemingly legitimate Web servers and not search for every other service. Assuming that there are 50 such machines out of 50,000 IP addresses and that all other conditions specified above remain the same, the first probe still yields a 0.1% chance of success, but after 50,000 probes, all 50 of the legitimate web services have been identified. This is less than one attempt from every IP address in a class B network, which one can easily simulate using the same deception technology as the defender with only a single machine which flexes IP addresses as needed for the attack.

The attack and countermeasure is a vicious circle, with no end to the process. Taking it further, with this level of deception, suppose that the defender correlated IP addresses of historical intelligence probes and, after 10 tries from different IP addresses in a class C network or 100 tries from different IP addresses in a class B network, switched the entire networks to deceptions. In this case, the probe above would be limited to 100 tries per class B network, so that in order for the attacker to identify all 50 Web servers, it would require the use of 50,000 IP addresses in 5,000 class C networks or 500 class B networks.

After determining that this sort of deception is in place, a stronger intelligence effort from the attacker would not concentrate on random probing. Instead, the effort might be on a more fruitful and more expensive process, such as following existing 'known-good' paths or deterministic paths into the infrastructure, planting insiders who can probe with far more knowledge, etc. Countering these intelligence attempts requires different deception advancement and the process goes on.

Enhancing the Quality of Deceptions

In any deceptive system, the quality of deception needs to be high to make it hard for an attacker to differentiate a legitimate service from a deception. Some of the areas that need enhancements are:

1. Traffic analysis should not be able to detect deceptions, so produce simulated traffic instead. There are two ways to simulate the traffic. One way is to reproduce the real traffic within the deception systems, either in real-time or in a replay mode. This makes one system appear to be another to the point where, all access attempts are duplicated in the deception. A person within the deception network can break into a system, sniff traffic, and use the traffic to break into other systems in the deception network, without getting access to the real network, and with tracking of all of their activities. The second method is to generate traffic from remote sites so that there is traffic that can be analyzed and exploited by an attacker.

2. As the normal network changes over time, the deception network should also be reconfigured to emulate the network changes over time. The real networks change over time, and if the deception is static, it will reduce to a passive intelligence in long run. One way to mitigate this is by changing the deception network over time as a real network does. The deception features should reflect the real-system characteristics as closely as possible. For example, if office computers are turned off at the end of the day, deception computers should simulate turned off machines in similar time frames. This can be done by following the behaviors of legitimate machines in near-real-time or by creating algorithms that do this [6]. Some care is needed in that holidays, weekends, and special situations must be accounted for or any decent intelligence organization will find the deceptions quickly.

3. Organizational deceptions should be created in a way that supports the technical deceptions. If the organization provides access to information about individuals or systems, then the deception must also reflect this information. For example, if the DNS in an organization includes details of individual system owners and locations, then false owners and the locations to go along with the DNS tables or deceptions will be needed.

Similarly, the fictitious people and locations one creates will need fictitious salaries, budgets, personnel records, personal problems, and phone numbers with people to answer them, and so on. The best move is to remove the intelligence information in places like the DNS system so as to mitigate the intelligence threat at low cost while also reducing the value of these services for other forms of intelligence. A negative side effect of this is that it has an impact on network maintenance and thus a method has to be devised to allow authorized people to access this information under appropriate circumstances and with some sort of detection of attempts to abuse this access. Of course this extends the number of people who may be able to tell deceptions from real systems, and such a scheme should also be distributed to the extent possible so that even the most trusted parties only know how to tell deceptions from real systems in their own sphere of influence, and then only under supervision. Clearly, this is one of the riskiest and most complex areas of undertaking in deception, and for that reason, the alternatives of less information available to more people are far more viable than those which lead to deceptions of enormous scale. One must pick where the scale is worthwhile in exchange for some intelligence advantage and balance the advantages of the availability of information with the risks it brings.

4. Instead of finite state machines as are used in DTK one may use real systems for deceptions. The use of real systems for deceptions is a method by which we replace the simple finite state machines of DTK with live systems of the proper type. In other words, the deceptions are precisely what one asserts them to be, except of course that they do not have the same content or actual functions as the machines they are designed to emulate. For that, deceptions are made by making the DTK-like elements of normal systems also behave through replacement systems.

To provide the means for far higher quality in deceptions, the technique shown in Figure 5.7 uses multiple address translation at a relatively very low cost.

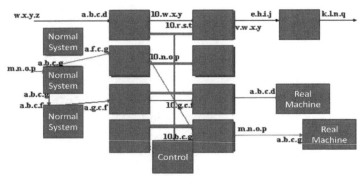

Figure 5.7 Multiple address translations.

The basic notion is to redirect selective services, e.g., a proxy service—where DTK used to handle the service directly—but rather than a simple proxy service, this service does address translation so that the same source and destination addresses remain in the deception system as were in place in the original system. In Figure 5.7, a user from m.n.o.p enters at the interface to a.b.c.g and is redirected through a sequence of address translations to a.f.c.g which transmits from 10.n.o.p toward 10.g.c.f and finally, the last translation has the packet going from the deception version of m.n.o.p to the deception version of a.b.c.g. Another example in Figure 5.8 has more than two address translations (i.e., w.x.y.z eventually becomes k.l.n.q).

This mechanism can be used for a wide variety of purposes, including flexing the translation mechanisms over time for diffusion of data through multi-hop anonymizer services, tunneling traffic through intervening infrastructures, and creating multiple deceptions based on source and destinations. In the case of enhanced deception quality, we can use this method to associate deception services to the same machine type and configuration as the real services thus dramatically increasing the realism associated with the deception.

Figures 5.8 and 5.9 show the advancement from DTK to deception wall (D-Wall) I and II respectively. DTK has deceptions that are slightly more than real vulnerabilities. D-Wall I had far more deceptions as compared to real vulnerabilities. So the seach space is much larger in D-Wall I. However, the deceptions may be of low quality. In case of D-Wall II, though the deceptions are slightly more than real vulnerabilities, their quality is high.

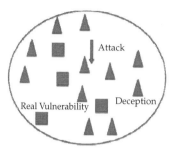

DTK: Deceptions are slightly
more than Real vulnerabilities

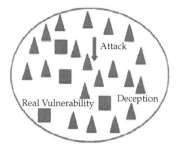

D-Wall: Very high concentration of
deceptions, larger search space

Figure 5.8 From Deception toolkit to deception wall—I.

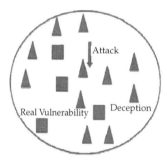

D-Wall (2): Deceptions are slightly
more than Real vulnerabilities, but
Quality high

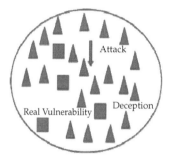

D-Wall (1): Very high concentration of
deceptions of low quality, larger search
space

Figure 5.9 From DTK to DWall—II.

5.3 SUMMARY

Static honeypot is fixed in location and number in a network. It can enhance the security in two ways. It can act as a Detection System or could be sued to deceive the attacker. A layered honeypot framework Japonica presents Honeypot as a Detection System. This framework consists of seven components arranged in four layers that interact with each other to defend the attacks in network. For using Honeypot as a Deception System, the techniques face the challenge to make high quality deception against technical intelligence efforts of the attackers. The combination of the Honeypot Model as Detection System with a more advanced approach to deception can dramatically increase the security in the network.

EXERCISES

Short Answer type Questions:

1. How can honeypot act as deception systems?
2. How is honeypot used as detection system?
3. Can a honeypot replace an IDS?
4. Bring out the advantages and disadvantages of increasing deception.
5. What are the goals and requirements of Japonica?
6. What is the advantage of having layered structure in Japonica?
7. Give significance of each of the four layers of Japonica. What do you mean by threat level and threshold in layer 2 of Japonica?
8. Is there any situation where thresholds be made dynamic like threat levels? Explain.

Long Answer type Questions:

1. Explain the two phases of conversion from Deception Toolkit to Deception Wall.
2. Explain the multiple address translation schemes to increase deception. Give example.
3. How can you enhance the quality of deception? Explain.
4. How can you increase the size of the search space and the sparsity of Real Vulnerabilities?
5. Contrast between early honeypot systems and deception spread among normal systems. Which has more value and why?
6. Discuss in details the role of each component of the four layers of Japonica framework and their interaction.

REFERENCES

[1] H. Artail, H. Safa, M. Sraj, I. Kuwatly, and Z.-A. Masri, "A hybrid honeypot framework for improving intrusion detection systems in protecting organisational networks," *Computers & Security,* vol. 25, pp. Pages 274–288. June 2006.

[2] L. Teo, Y.A. Sun, and G.J. Ahn, "Defeating Internet attacks using risk awareness and active honeypots," in *Second IEEE International Information Assurance Workshop (IWIA'04)*, Charlotte, North Carolina 2004, pp. 155–168.

[3] C. E. Shannon, "A mathematical theory of communication," *ACM SIGMOBILE Mobile Computing and Communications Review,* vol. 5, p. 55, 2001.

[4] N.C. Rowe, "Measuring the effectiveness of honeypot counter-counterdeception," in *Proceedings of the 39th Annual Hawaii International Conference on System Sciences*, Hawaii, 2006, pp. 129c–129c.

[5] F.C. Associates, "A Mathematical Structure of Simple Defensive Network Deceptions," in *Strategic Security Intelligence*, p. http://all.net/journal/deception/mathdeception/mathdeception.html.

[6] R. Sekhar, T. bowen, and M. Segal, "On preventing intrusions by process behavior monitoring," in *Proceedings of the 1st conference on Workshop on Intrusion Detection and Network Monitoring—vol. 1 table of contents,* 1999 p. 4.

6

Virtual Honeypots

This chapter describes virtual honeypots and data capture on virtual honeypots. It then draws out the difference between raw disks and virtual disks and further elaborates the more complex concept of virtual Honeynets. Finally it presents a practical case study of attack capture using virtual honeypot VMWare.

Virtual Honeypot [1] is a software program that is designed to appear to be a real functioning system but is actually a decoy built specifically to be probed and attacked by malicious users. In contrast to a honeypot, that is typically a hardware device that lures users into its trap, a virtual honeypot uses software to emulate a network.

The difference between a real and a virtual honeypot lies in the fact that a virtual honeypot uses application software to create a new, separate operating system environment. The virtual host actually uses or shares that same hardware as the physical OS does. Instead of using different hardware for each host, many different virtual servers may be contained on one piece of hardware. A First Generation (GenI) Honeypot/Honeynet uses separate hardware for each network host while Second Generation (GenII) Honeypots/ Honeynets uses virtual software to condense the environment down to one piece of hardware.

6.1 VIRTUAL HONEYPOT: VMWARE WORKSTATION

VMware [2] is a well-established virtualization option for Linux and Windows platforms. Some advantages of using VMware Workstation include:

- **Wide range of operating system support:** It runs a variety of operating systems within the virtual environment (called GuestOSs), including Linux, Solaris, Windows and FreeBSD honeypots.
- **Networking options:** Workstation provides two ways to handle networking. The first is bridged, which is useful for hybrid virtual honeynet networks because it lets a honeypot use the computer's card and appear to be any other host on the honeynet network. The second option is host-only networking, this is good for self-contained virtual honeynets because one is able to better control traffic with a firewall.
- **VMware Workstation creates an image of each Guest Operating System:** These images are simply a file, making them highly portable. This means one can transfer them to other computers. To restore a honeypot to its original condition, one can just copy a backup into its place.
- **Ability to mount VMware virtual disk images:** One is able to mount a VMware image just like one would a drive using vmware-mount.pl.
- **Easy to use:** VMware [3] Workstation comes with a graphical interface (both Windows and Linux) that makes installing, configuring, and running the operating systems very simple.
- **Support:** As a commercial product, VMware Workstation comes with support, upgrades, and patches.

Some disadvantages of VMware Workstation include:

- **Resource requirements:** VMware Workstation must run under an X (or GUI) environment and each virtual machine will need its own window. So on top of the memory that is allocated for the GuestOSs (the OS that is installed on a virtual machine), there is the overhead of the GUI system.
- **Limited number of GuestOSs:** With VMware one can only run a small amount of virtual machines (approximately 1-4). This might make for a small honeynet.

Closed Source: since VMware is closed source, users can't really make any custom adjustments.

6.2 DATA CAPTURE ON VIRTUAL HONEYPOTS

Data capture is the capturing of all of the blackhat's activities. It is these activities that are then analyzed to learn the tools, tactics, and motives of the blackhat community. The challenge is to capture as much data as possible, without the blackhat knowing their every action is captured. This is done with as few modifications as possible, if any, to the honeypots. Also, data captured cannot be stored on locally on the honeypot. Information stored locally can potentially be detected by the blackhat, alerting them the system is a Honeynet. The stored data can also be lost or destroyed. Not only one has to capture the blackhats every move without them knowing, but the information has to be stored remotely.

The first method for data capture is to have all system logs not only log locally, but to a remote log server. For Unix systems and most network devices, this is simply done by adding an entry for a remote syslog server in the configuration file. For Window based systems there are third party applications that will remotely log information. Also, system logs can be written on the remote log server. NT requires a third party software to have the capability to write system information to syslog, but it can write it to a network file system. This way, critical system information such as process activity, system connections, and attempted exploits are safely copied to a remote system. Any attempt to hide the use of a remote syslog server should not be made. If the blackhat detects this, the worse they can do is disable syslogd (which is standard behavior for most blackhats). This means there will no longer be continued logs, however one will at least have information on how they gained access and from where.

i) More advanced blackhats will attempt to compromise the remote syslog server in an attempt to cover their tracks. This is exactly what is required. The syslog server is normally a far more secured system. This means for a blackhat to successfully take control of such a system they will have to use more advanced techniques, which one will capture and learn from. If the syslog server is compromised, nothing is lost. Yes, the blackhat can gain control of the system and wipe the logs. IDS system that is on the network passively captures and records all of the logging activity that happened on the network. In reality, the IDS system acts as a second remote log system, as it passively captures all the network data.

ii) A second method to capturing system data is to modify the system to capture keystrokes and screenshots and remotely forward that data.

Specific requirements for data capture are:

i) No Honeynet captured data should be stored locally on the honeypot. (Data logged on honeypots is assumed to be unreliable, and may be modified by intruders.) Honeynet captured data is any logging or information capture associated with activity within a Honeynet environment.

ii) No data pollution can contaminate the Honeynet, invalidating data capture. Data pollution is any activity that is non-standard to the environment. An example would be a non blackhat testing a tool by attacking a honeypot.

iii) The following activity must be captured and archived:
   a) Network Activity
   b) System Activity
   c) Application Activity
   d) User Activity

iv) The ability to remotely view this activity in real time.

v) The automated archiving of this data for future analysis.

vi) Maintain a standardized log of every honeypot deployed.

vii) Maintain a standardized, detailed write-up of every honeypot compromised.

viii) Resources used to capture data must be secured against compromise to protect the integrity of the data.

The information listed above outlines the requirements for Data Capture logging within Honeynets and breaks down to the following three types of monitoring:

i) Firewall logging of both inbound/outbound connections,

ii) Network Intrusion Detection logging of all traffic, and

iii) Host based modifications to log both system and user activity.

In honeypot data capturing host based monitoring is used. Even with the most stealthy host-based modifications, such as the loadable kernel modules for keystroke logging, the fact is that any changes made to the honeypot can potentially be identified by the attacker. The challenge is to monitor activity within a virtual honeypot system without making any logging modifications to the honeypot system itself.

One of the distinct advantages to using virtual honeypots over normal separate honeypot systems, is the fact that the virtual Guest honeypot system is actually running inside the Host OS system. With this configuration, one is able to access (monitor) the files which make up the virtual honeypot system without having to use the network to do so. There are two different ways to implement a VMware Guest OS system onto a Host OS system—either using "Raw Disks" or "Virtual Disks"[4]. Both implementations allow the Host OS system to access the Guest OS files. It is this ability which has been utilized to monitor Guest OS honeypots.

6.3 RAW DISKS AND VIRTUAL DISKS

Existing Physical (Raw) Disk: An existing physical disk or raw disk [2] is a partition on a physical IDE or SCSI drive connected to the host computer. Raw disks are used if one wants Workstation to run one or more guest operating systems from existing disk partitions. Raw disks may be set up on both IDE and SCSI devices.

Virtual Disk: A virtual disk [2] is a file on the host file system that contains all data stored in the virtual IDE or SCSI drives visible to the guest operating system. A virtual disk can be created on any type of disk (IDE, SCSI, etc.) and any type of file system (E2FS, FAT, FAT32, NTFS, etc.) supported by the host operating system. The virtual disk can also be created on a removable disk drive or placed on a network file server. The New Virtual Machine Wizard will place the virtual disk in the virtual machine directory specified. Virtual disks can be as large as 128GB for IDE virtual hard disks and 256GB for

SCSI virtual hard disks. Workstation creates a file for each 2000MB of virtual disk capacity. The actual files used by the virtual disk start out small and grow to the maximum size as needed. A key advantage of virtual disks is their portability. Because the virtual disks are stored as files on the host machine or a remote computer, they can be easily moved to a new location on the same computer or to a different computer.

Virtual disks are a set of files that VMware presents as a "real" hard drive to the guest operating system, raw disk partitions are a "real" partition on a "real" hard drive that the guest operating system is given access to. There are several advantages and disadvantages to either approach. If the purpose is quick research or primarily as an early warning device and one does not plan to prosecute it is acceptable to use virtual disks. They allow for easy copying and recreation of a honeypot once it has been compromised and are the simplest to install. However if the guest operating system is sufficiently damaged one will not be able to access it very easily, and since it uses a custom file format one will not be able to examine it easily with common forensics tools. The major advantage of virtual disks is convenience, however one will lose much of any ability to perform forensics.

For deeper forensics, and especially if one plans to prosecute the best option is to use raw disk partitions. Physically separate hard drive from the host operating system's hard drive eases partitioning and make potential cross contamination less likely. Although it is unlikely that the attacker will be able to break into a properly secured VMware host machine it is possible.

There are some advantages to using the virtual disks. First of all, it is easier to install the VMware Guest OS host since one does not have to bother with repartitioning the Host system. These files are also portable so that one can transfer them to other system rather easily.

6.4 VIRTUAL HONEYNET

Virtual Honeynets [5] represent a relatively new field for Honeynets. The concept is to virtually run an entire Honeynet on a single, physical system. The purpose of this is to make Honeynets a cheaper solution that is easier to manage. Instead of investing in large amounts of hardware, all of the hardware requirements are combined onto a single system.

Virtual Honeynets combine all the elements of a Honeynet onto one physical system. Not only are all three requirements of Data Control, Data Capture, and Data Collection met, but the actual honeypots themselves run on the single system. Virtual Honeynets do not represent a specific architecture; they can support either GenI or GenII technologies. Instead, virtual Honeynets represent one option for deploying these architectures. The honeypots are actual operating systems. Nothing is emulated. The advantage here is one of cost and efficiency. It is much cheaper to use a single system to run all the elements of a Honeynet, and it is much easier to deploy and maintain.

There are several different methods to deploying virtual Honeynets. The options are based on what type of emulation solution you select for virtual Honeynet. Each solution has its advantages and disadvantages. Two virtual Honeynet options are VMware and User Mode Linux. VMware is a commercial, supported solution designed to run multiple operating systems at the same time. VMware runs only on the Intel architecture, so only Intel-based operating systems work with VMware. For example, a GenI virtual Honeynet could be created using VMware, running Linux, Windows 2000, Windows XP, and Solaris X86 (a version of Solaris designed for the Intel architecture). User Mode Linux (commonly called UML) is an Open Source solution with similar functionality. However, UML currently is limited to the Linux operating system.

While virtual Honeynets have several advantages, their main disadvantage is that you are limited to the operating systems and architectures supported by the emulation software. For example, neither VMware nor UML can support Sun Microsystems Sparc architecture or the Cisco IOS. Also, there may be additional risk of fingerprinting with the virtual solutions.

6.5 CASE STUDY

Gathering information about enemy is important. By knowing attack strategies, countermeasures can be taken and vulnerabilities can be fixed. To gather as much information as possible is one main goal of a honeypot. Generally, such information gathering should be done silently, without alarming an attacker. All the gathered information can be used on productive systems to prevent attacks.

The case study presents a technique to monitor both user and system activity within a VMware [2] Guest OS honeypot system. The system activities include both the creation and content of new files and directories, including rootkits, live sniffer logs, etc. Different methods to identify and extract information and files directly from the live virtual guest OS [6] without altering the pre configuration of the honeypot host have been implemented. These methods also present an alternative to conducting the normal live forensic audit of the honeypot system, where data contamination such as altering often occurs. Once this data has been extracted from the live honeypot it can then be analyzed using standard forensic techniques and tools.

Biggest challenges for the Honeypot have been their implementation and finding methods of monitoring that will, hopefully, be undetectable by the intruder [7]. In the past, monitoring has included network monitoring, remote log collection, key-stroke capturing, etc.

Some initial tests using SWATCH for monitoring did not work effectively. This was due to SWATCH's reliance on PERL's TAIL function to search for data. Since the Honeypot virtual disk files could change content anywhere within the file, and not just at the end, SWATCH was not a reliable method of monitoring. Monitoring techniques also required alterations in honeypot and re-configurations of host OS which were sometimes impossible to obtain.

The case study presented implements a honeypot VMWare that can be deployed with existing configuration of host OS and monitor the entire data captured by Honeypot without alterations in the Honeypot. The study is done under two parts:

1. Implementation of Virtual Honeypot VMware that creates large number of honeypot systems in one machine.
2. Data capture, monitoring the captured data and its analysis using forensic tool xtail.

The testing is conducted on a closed network. The network consists of one Windows XP Professional machine, which was the VMware

Host OS machine. There is one Linux Fedora Guest OS machines on the Windows XP Professional machine. There is also a SUN Solaris Workstation on the test network. Since the test is conducted on a closed network, the possibility for outside network interference is minimized. The test environment used for the tests is:

Hardware:

- 1—IBM desktop system
- 1—HP—Laptop system
  These systems would function as the VMware Host OS systems for testing.
- SUN Solaris Workstation.
  It has been used to test remote system accessing the VMware honeypot system.

Software:

- Windows XP professional OS. This system functions as the VMWare Host OS system for testing. Figure 6.1 shows the system settings.

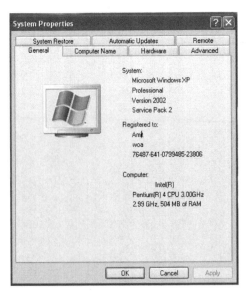

Figure 6.1 Windows XP Professional System Settings.

- Linux Fedora
- VMware
- Cygwin
- Xtail

Cygwin Version 2.125.2.10

Cygwin [8] is a UNIX environment, developed by Red Hat, for Windows. It consists of two parts:

- A DLL (cygwin1.dll) that acts as a UNIX emulation layer providing substantial UNIX API functionality.
- A collection of tools, ported from UNIX, which provide UNIX/ Linux look and feel.

The Cygwin DLL works with all non-beta, ix86 versions of Windows. Figure 6.2 shows Cygwin terminal screen.

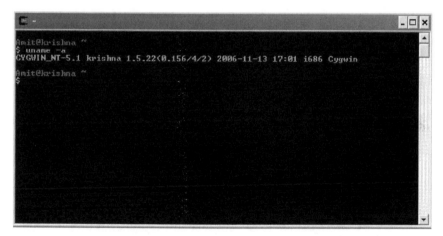

Figure 6.2 Cygwin Terminal Screen.

VMware Windows Workstation—V.5.5.2-build 29772

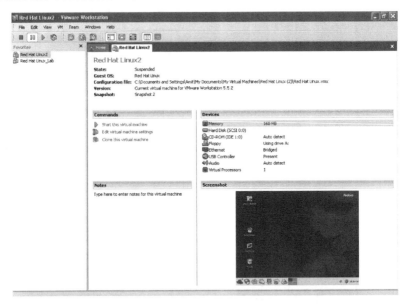

Figure 6.3 VMware Workstation Application Screen.

Linux Fedora

This is the VMware Guest OS-Kernel 2.2.14-5.0 i686-No Patches. Figure 6.4 shows Windows XP Professional machine with the Linux VMWare Guest OS running inside it.

Figure 6.4 Windows XP Host OS with VMware Linux Fedora Guest OS.

Laboratory Test Network

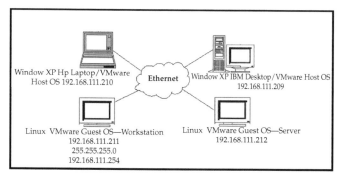

Figure 6.5 Test Network.

Configuration Details

VMware Install [2]

The key configurations during the installation process to create a honeypot environment that will assist with capturing both system and user activity within the VMware honeypot Guest OS system are discussed.

Amount of memory that should be allocated to this virtual machine: The default amount of memory that is allocated to a new virtual machine created with VMware Workstation 5 depends on:

i) the guest operating system configured for the virtual machine.

ii) the host operating system (Windows or Linux).

iii) the amount of physical memory installed in the host machine.

iv) the memory limit for all virtual machines (referred to as reserved memory) that is set in the Settings > Preferences > Memory configuration screen.

Following steps were followed to configure the memory for a virtual machine:

i) Select File > Open and open the virtual machine configuration file (.vmx) to be modified.

ii) Select Settings > Configuration Editor.

iii) Click the Hardware panel.

iv) Select Memory from the list of devices.

v) In the "Guest size (MB)" field, enter the total memory (in megabytes) that has to be assigned to this virtual machine. Memory size is restricted to multiples of 4 megabytes.

vi) Click OK.

Figure shows the memory configuration screen:

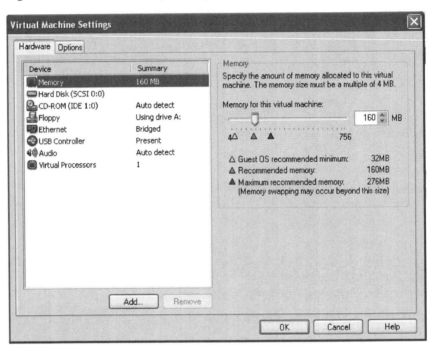

Figure 6.6 VMware Memory Configuration Screen.

Virtual honeypot system has been configured to use the least amount of memory as possible. This will most likely cause the Guest OS system to write most of its data that would normally be held within memory to the local swap file. When this data is written to the swap file, there will be a greater chance of capturing this data.

Configuring the Disk Mode for a Virtual Machine

There are three different disk modes that can be enabled for both raw and virtual disks:

Persistent

This mode is the default. With this mode, all data written to a persistent disk are immediately and permanently written to the disk. As a result, the disk behaves as a conventional disk drive on a real computer.

Nonpersistent

Changes to nonpersistent disks are not saved during the Workstation session and are lost at the end of the session (that is, when the virtual machine is powered off or reset). Nonpersistent disks are convenient for people who always want to start with a virtual machine in the same state. Example uses include providing known environments for software test and technical support users as well as doing demonstrations of software. This type of disk mode could also be useful for a malware analysis test platform where all changes to the system can be removed when the system is powered off.

Undoable

This is the disk mode that have been configured on Guest OS Honeypot system to use. Undoable mode lets one decide at the end of a working session whether you want to keep or discard the changes made during that session. This is especially useful for experimenting with new configurations or unfamiliar software. This disk mode is extremely useful for honeypot deployments.

Undoable disks are similar to nonpersistent disks in that writes to the disk are stored in a file called a redo log. An undoable disk, unlike a nonpersistent disk, gives the option later of permanently applying the changes saved in the redo log, so they become part of the main disk. While the Workstation session is running, disk blocks that have been modified and written to the redo log are read from there instead of the disk. Any disk type can be used in undoable mode.

With the implementation of the undoable virtual disks, one is able to have a baseline of verified, freshly installed system with the normal VMware disk files and separate any changes made to the system, presumably by attackers, into the redo log files. This identifies and monitors all of the changes made to honeypot system by analyzing the redo logs instead of the normal virtual disk files.

In VMware 5 Workstation snapshot is used to implement Undoable mode. The snapshot captures the entire state of the virtual machine at the time of the snapshot. This includes:

i) The state of all the virtual machine's disks.

ii) The contents of the virtual machine's memory.

iii) The virtual machine settings.

When one reverts to the snapshot, one return all these items to the state they were in at the time the snapshot was taken.

The Snapshot and Undoable Mode

Snapshot has been used to achieve results equivalent to that of Undoable mode. A snapshot is taken when working session begins. To discard all work done during the session, revert to the snapshot. To commit the work done during the session, take a new snapshot at the end of the working session. To keep the work done during a session without committing it, leave the original snapshot unchanged. Figure illustrate this.

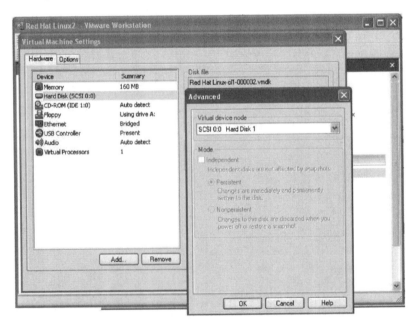

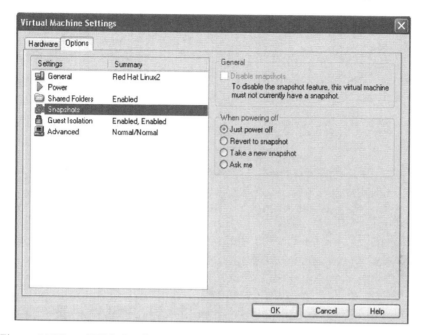

Figure 6.7 Virtual Disk Configuration Screens.

Key Concept—Undoable Disks

Undoable virtual disks are the key factor for effective VMware honeypot monitoring.

Before conducting forensic analysis of compomised systems, the normal process is to create exact replicas of the entire system. This is normally achieved by using tools to create a bit by bit copy of each partition on the system. This is necessary because:

i) Integrity of the Data—By using tools such as MD5, Forensic Analysts can validate the integrity of the data. This issue is crucial if the evidence will ever be used in future legal actions. If this data will be used to prosecute an attacker, then an exact replica of the system must be made.

ii) Create Data Images for Forensic Examination—In order to have data to examine, one has to create replicas. This will allow one to have data to work with for the examination process while not accessing and/or altering the original.

The down side of this process is that the resulting images used for forensic analysis are extremely large, i.e., many GBs of data. Forensic Analysts must comb through vast amounts of data and search for attack data. The problem with the forensic analysis of these huge files is data overload. Using the VMware Undoable disks actually helps to alleviate the issue of data overload by narrowing the scope of the interest to only new data.

Working of Undoable Disk Mode

Using the Undoable virtual disks when monitoring VMware honeypots allows the WhiteHats to focus in on ONLY the new system data introduced to the honeypot system. Any new system activity stored in the VMware REDO log files is suspicious by definition. By focusing in on only the REDO log file data instead of the normal OS files, results is a significant decrease in the amount of data to analyze. Xtail is the tool that has been used to monitor the REDO logs.

Xtail Overview

Xtail [2] is a utility that will watch the growth of one or more files in a directory and print the content to the screen. Here is the information from the xtail MAN page:

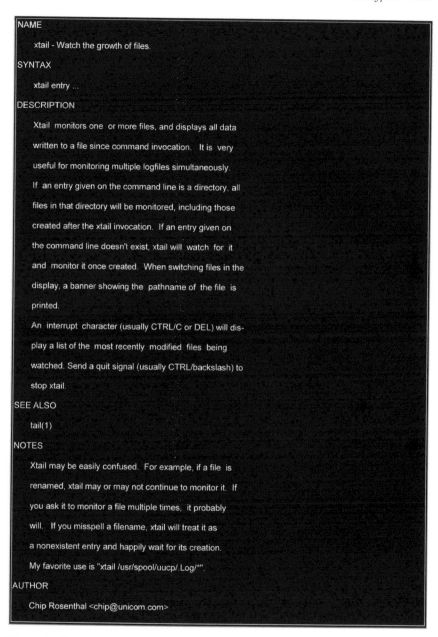

```
NAME

    xtail - Watch the growth of files.

SYNTAX

    xtail entry ...

DESCRIPTION

    Xtail  monitors one  or more files, and displays all data

    written to a file since command invocation.  It is  very

    useful for monitoring multiple logfiles simultaneously.

    If  an entry given on the command line is a directory, all

    files in that directory will be monitored, including those

    created after the xtail invocation.  If an entry given on

    the command line doesn't exist, xtail will  watch  for  it

    and  monitor it once created.  When switching files in the

    display, a banner showing the  pathname of  the file  is

    printed.

    An  interrupt  character (usually CTRL/C or DEL) will dis-

    play a list of the  most recently  modified  files  being

    watched. Send a quit signal (usually CTRL/backslash) to

    stop xtail.

SEE ALSO

    tail(1)

NOTES

    Xtail may be easily confused.  For example, if a file  is

    renamed, xtail may or may not continue to monitor it.  If

    you ask it to monitor a file multiple times,  it probably

    will.  If you misspell a filename, xtail will treat it as

    a nonexistent entry and happily wait for its creation.

    My favorite use is "xtail /usr/spool/uucp/.Log/*".

AUTHOR

    Chip Rosenthal <chip@unicom.com>
```

Figure 6.8 Xtail MAN page.

Even though xtail says that it is similar in functionality to running a "tail -f filename", command, it is actually quite different. Whereas tail will only monitor any changes appended to the end of a file, xtail actually monitors the entire file. When xtail is invoked with the following command:

```
# xtail /var/log/messages
```

Figure 6.9 The Xtail command.

It will monitor the /var/log/messages file for any changes, this includes both new data and deletions. Xtail uses two different parameters within its C looping function to determine if a file that it is monitoring has changed.

- The size of the file
- The mtime of the file

Source code from the xtal.c file which show how the modifications are handled:

```
/*
  * Go through all of the files looking for changes.
  */
Dprintf(stderr, ">>> checking files list (%s)\n",
    (open_files_only ? "open files only" : "all files"));
for (i = 0 ; i < List_file->num_entries ; ++i) {
    entryp = List_file->list[i];
    already_open = (entryp->fd > 0) ;


    /*
      * Get the status of this file.
      */
    switch (stat_entry(List_file, i, &sbuf)) {
    case ENTRY_FILE:  /* got status OK  */
break;
}



    /*
      * If the file isn't already open, then do so.
      * It is important toe call "fixup_open_files()"
      *   at the end of the loop to make sure too many files don't
      *   stay opened.
      */
    if (!already_open && open_entry(List_file, i) != 0) {
--i;
continue;
```

This code shows how xtail continuously searches through the list of files looking for any changes and reporting this information to standard output.

Building Xtail

Before xtail could be used to monitor the VMware Guest OS REDO files, xtail source code was compiled to create a Cygwin executable binary. The xtail-2.1.tar file was moved into the VMware Guest OS directory and following steps were executed to compile xtail.

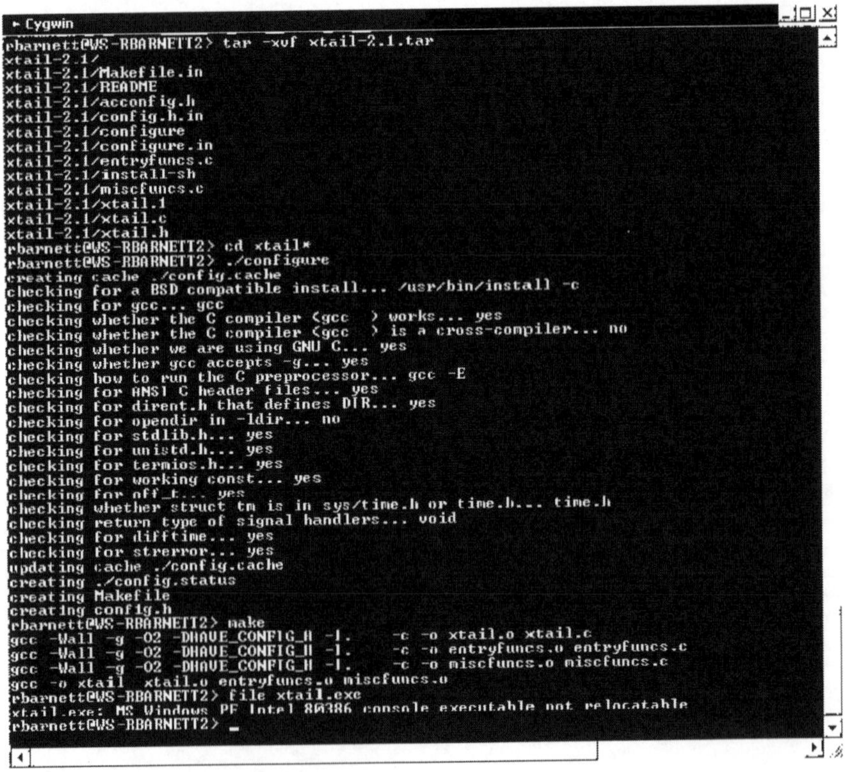

```
► Cygwin                                                        _ | □ | x|
rbarnett@WS-RBARNETT2> tar -xvf xtail-2.1.tar
xtail-2.1/
xtail-2.1/Makefile.in
xtail-2.1/README
xtail-2.1/acconfig.h
xtail-2.1/config.h.in
xtail-2.1/configure
xtail-2.1/configure.in
xtail-2.1/entryfuncs.c
xtail-2.1/install-sh
xtail-2.1/miscfuncs.c
xtail-2.1/xtail.1
xtail-2.1/xtail.c
xtail-2.1/xtail.h
rbarnett@WS-RBARNETT2> cd xtail*
rbarnett@WS-RBARNETT2> ./configure
creating cache ./config.cache
checking for a BSD compatible install... /usr/bin/install -c
checking for gcc... gcc
checking whether the C compiler (gcc ) works... yes
checking whether the C compiler (gcc ) is a cross-compiler... no
checking whether we are using GNU C... yes
checking whether gcc accepts -g... yes
checking how to run the C preprocessor... gcc -E
checking for ANSI C header files... yes
checking for dirent.h that defines DIR... yes
checking for opendir in -ldir... no
checking for stdlib.h... yes
checking for unistd.h... yes
checking for termios.h... yes
checking for working const... yes
checking for off_t... yes
checking whether struct tm is in sys/time.h or time.h... time.h
checking return type of signal handlers... void
checking for difftime... yes
checking for strerror... yes
updating cache ./config.cache
creating ./config.status
creating Makefile
creating config.h
rbarnett@WS-RBARNETT2> make
gcc -Wall -g -O2 -DHAVE_CONFIG_H -I.    -c -o xtail.o xtail.c
gcc -Wall -g -O2 -DHAVE_CONFIG_H -I.    -c -o entryfuncs.o entryfuncs.c
gcc -Wall -g -O2 -DHAVE_CONFIG_H -I.    -c -o miscfuncs.o miscfuncs.c
gcc -o xtail  xtail.o entryfuncs.o miscfuncs.o
rbarnett@WS-RBARNETT2> file xtail.exe
xtail.exe: MS Windows PE Intel 80386 console executable not relocatable
rbarnett@WS-RBARNETT2> _
```

Figure 6.10 Xtail Configuration.

The resulting /path/to/VMware/Linux/ directory contained the files. Figure 6.11 shows the same directory contents from both the VMware Host OS (Windows XP Professional) and the Cygwin Terminal window.

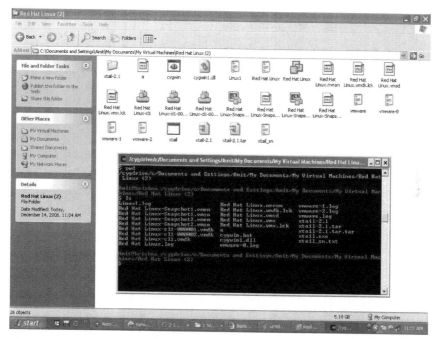

Figure 6.11 VMware Guest OS Directory Content with Cygwin Terminal.

Forensic Preparation

Before deployment the VMware Honeypot system, following forensic steps were conducted to aid in Incident Response steps.

i) Create a replica of the VMware honeypot system

Honeypot system named Red Hat Linux(2) was created and then an exact replica of this VMware host was created. This replica will be used as Forensic Test Lab server. To accomplish this task, all of the VMWare Guest OS files were copied into a new directory and renamed Red Hat Linux_Lab. This replica of the original Red Hat Linux(2) VMware allowed to move the updated changes (attacker modifications) from the live honeypot system to the replica server for analysis. After copying and renaming the files, the/path/to/ VMware/Red Hat Linux_Lab/linux.vmx file was updated. This file holds the configuration information about the VMware host.

```
config.version = "6"

virtualHW.version = 2

displayName = "Red Hat Linux_Lab"

scsi0:0.mode = "undoable"

draw = "gdi"

guestOS = "linux"

ide1:0.present = TRUE

ide1:0.deviceType = "atapi-cdrom"

ide1:0.fileName = "auto detect"

scsi0:0.present = TRUE

scsi0:0.fileName = "C:\Documents and Settings\Amit\My Documents\My

Virtual Machines\Red Hat Linux_Lab\Linux1.vmdk"

scsi0.present = TRUE

ethernet0.present = TRUE
```

ii) If one fails to update this file, then the live honeypot virtual disks will be used instead of the Forensic Test VMware host's files. This would cause data pollution in the live honeypot system. However, the goal of this entire process is to monitor and extract changed data without modifying the original honeypot host.

iii) Create a baseline of all Honeypot system files.

Log into the VMware Red Hat Linux_Lab server and create an MD5 baseline of every file on the system. This baseline will be used to identify any changes made (attacker modifications) to the VMware Red Hat Linux(2) honeypot system. There are many different tools which can be used for this process such as Tripwire, AIDE, etc. Tripwire was installed and a database created from the linux tw.config file. Once this new tripwire database was created, both the linux tw.config file and the tripwire database—tw.db_amit-was copied onto the Windows XP VMware Host OS system. This database can be used later to verify data when new REDO log data is introduced from the Red Hat Linux(2) VMware honeypot into the VMware Red Hat Linux_Lab host.

Once these steps have been completed, Red Hat Linux_Lab VMware host was shut down. The system asked me if whether to commit the system changes which were stored in the Red Hat Linux_ Lab REDO files to disk. NO was selected to remove all information in the REDO logs and keep the Red Hat Linux_Lab virtual disks in the same state as before any of these Forensic Preparation steps were conducted.

Using Xtail to Monitor Live VMware REDO Logs

Red Hat Linux(2) VMware Guest honeypot system was started. Xtail should not monitor the REDO log files during this timeframe since the virtual disks will be performing normal system startup procedures. Once the VMware host is booted up into full interactive mode, then local REDO log files were monitored with xtail to identify any changes made to the system.

Since there will be multiple REDO files within the Red Hat Linux(2) VMware systems directory, xtail should monitor any file in this directory with the "*REDO" extension. These are the files that will contain all of the new system data. If all of the files in this directory are monitored, there will be data which one is not interested in, such as the data in the vmware.log file. This log file simply logs the VMware applications data such as when you type "Ctrl+Alt" to exit the VMware application to return to the Host system.

Due to the fact that the VMware REDO virtual disks will contain both ASCII text and Binary data, one does not want to use xtail to report on this data directly to the terminal screen. The techinque used actually takes the output from xtail and then pipes the data into the "strings -a" command. This will extract out only ASCII text. This data was then redirected to a log file called—Linux1.log and run this as a background process. This allowed to monitor the Linux1.log file by using the "tail -f Linux1.log" command. The the full command string to activate xtail to monitor the REDO is:

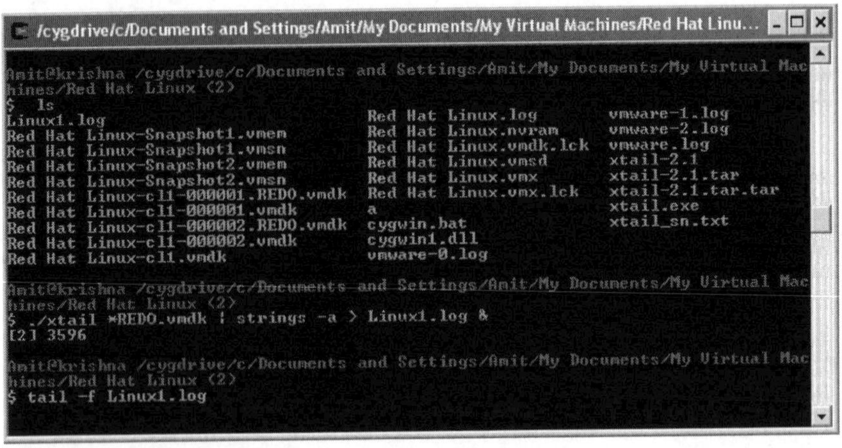

Figure 6.12 Starting Xtail to Monitor the REDO Logs.

Test Scenario and Results

The testing scenario for this VMware honeypot monitoring should achieve the following goals:

i) Capture all new data created in the VMware REDO virtual disk files.

ii) Provide the ability to monitor this data in real time with the use of xtail.

iii) Provide the ability to review the ASCII text of this data in either real time or in post-mortem log file analysis.

iv) Be able to conduct additional, common forensic analysis of the REDO information by introducing the data into the forensic VMware host—Red Hat Linux_Lab in future.

v) Finally, this technique should abide by the rules set forth by the Honeynet Project for Data Capture.

vi) Hide the monitoring from the attacker.

vii) Ensuring the integrity of the data that is captured.

The Honeypot Compromise

This test will simulate a typical honeypot compromise by running ADMsniff attack against the Red Hat Linux(2) VMWare honeypot system. The goal of this test from the attacker's perspective is to unzip and then download and install the ADMsniff and use it to sniff telnet sessions. The goal from the WhiteHat's perspective is to be able to effectively monitor all of this activity by using xtail to monitor the VMware REDO virtual disk file. Further capture the live network/attacker activity.

Figure 6.13 shows both the VMware Guest OS Linux Desktop honeypot system and the Windows 2000 Host OS system with the cygwin/Xtail applications running.

Figure 6.13 VMware Guest Linux OS with Cygwin/Xtail Application Monitoring.

Figure displays the data captured by Xtail when the file called ADMsniff.tar.gz was unzipped within the Red Hat Linux(2) Honeypot:

Figure 6.14 Xtail Data Capture During a Gunzip of the ADMSniffer Archive.

This last example screen shot shows how Xtail can capture live user/network activity. As shown in Figure, telnet was used to log into. VMware server on the right hand sideshows one logged into the host and greated with banner message. Back in the Xtail window on the left, Xtail captured this data.

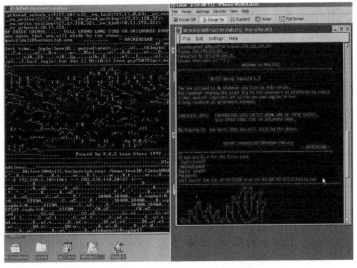

Figure 6.15 Xtail Captures a Telnet Session.

These figures give a visual representation of the Xtail monitoring setup between the Host OS and the Guest OS.

The goal of the study is to highlight a new method for monitoring virtual (VMware) honeypot systems.

The sophistication of the attacker's tools increases. This includes rootkit tools and techniques to identify hidden processes which are being used by WhiteHats to monitor their honeypots. Even if the WhiteHat efficiently hides processes that are running on the honeypot, the odds are that a skilled attacker could identify this monitoring. This is especially true if the attack installs a sniffer and detects data leaving the honeypot system. While it is true that encrypting this honeypot data leaving the host will protect the data, it will still tip off that attacker that there is a rogue process sending out data. This will most certainly prompt the attacker to act in a different manner than if they were unaware of the monitoring. One of the main advantages to the Xtail VMware monitoring is that the attacker data can both be identified and captured on the Host OS without modifying the Guest OS honeypot. This means that there are no scripts, files or processes to tip off the attacker.

6.6 SUMMARY

Honeypots and Honeynets have been replaced by their virtual counterparts. They provide an advantage of less hardware requirements and being easily manageable. They are also a cost effective mechanism and more efficient than deploying honeypots. Virtual disks are created on guest OS of real hard drive. The concept of virtual disks has been used to deploy multiple OS leading to creation of Honeynets. Finally use of Honeynet VMWare has been demonstrated to capture attacks.

EXERCISES

1. What is the difference between raw disk and physical disk?
2. Highlight the advantages of using virtual Honeynets.
3. What are the advantages of using VMware Workstation ?
4. Explain data capture on virtual honeypots.

REFERENCES

[1] F. Zhang, S. Zhou, Z. Qin, and J. Liu, "Honeypot: a supplemented active defense system for network security," in *Fourth International Conference on Parallel and Distributed Computing, Applications and Technologies*, Hamburg, 2003, pp. 231–235.

[2] http://www.vmware.com.

[3] T. Holz and F. Raynal, "Detecting honeypots and other suspicious environments," in *The Sixth Annual IEEE SMC Information Assurance Workshop*, 2005, pp. 29-36.

[4] E.K. Lee and C.A. Thekkath, "Petal: Distributed virtual disks," *ACM SIGOPS Operating Systems Review*, vol. 30, pp. 84–92, 1996.

[5] "Know Your Enemy: Building Virtual Honeynets," in *The Honeynet Project*, 2002, http://www.symantec.com/connect/articles/know-your-enemy-building-virtual-honeynets.

[6] K. Keahey, K. Doering, and I. Foster, "From sandbox to playground: Dynamic virtual environments in the grid," in *Grid Computing, 2004. Proceedings. Fifth IEEE/ACM International Workshop on*, 2004, pp. 34–42.

[7] N. Provos, "A virtual honeypot framework," in *Proceedings of the 13th conference on USENIX Security Symposium* San Diego, CA 2004, p. 1.

[8] J.Racine, "Cygwin Tool." Journal of Applied Economics vol. 15, 2000, pp. 331–341.

7

Dynamic Honeypots

This chapter briefly introduces the notion of dynamic honeypots and starts with the motivation behind making honeypots dynamic. It then presents how dynamic honeypots work. Dynamic honeypot approach integrates passive and active probing and virtual honeypot approach. Further we discuss the design of dynamic honeypot. Finally, this chapter presents the construction of dynamic honeypot and concludes with the advantages of dynamic honeypots over static honeypots.

7.1 ISSUES WITH STATIC HONEYPOTS

One of the biggest challenges with most security technologies, including honeypots, is configuring them [1]. Everything from encryption keys to firewall rules require a human to analyze the problem, come up with a solution, then configure and implement that solution. Adding more complexity, the work is not done once the technology is implemented. Once you have that 'bad boy' deployed, it can require daily updating and troubleshooting.

Honeypots face the same challenge, regardless of what type of honeypot one is building. From something as simple as Specter to something as advanced as Honeynets, configuration is still required. With honeypots, one of the configuration issues is *what will your honeypot look like*? Do you want it to appear as a Solaris system, Windows, Linux, or perhaps Cisco IOS? You want to ensure you are running the same operating systems that are deployed within your organization, so the honeypots can blend in with your environment.

Once you have determined the OS, what services do you want to run: web, mail, or perhaps file sharing? Failing to run the correct services means missed probed or attacks. However, monitoring the wrong services can be just as harmful, as the wrong service can be a dead giveaway for a honeypot. Having a Linux honeypot emulate Sub7 or NetBus services would look quite odd. Having a Windows honeypot emulate the sadmind, dtscpd, or rpc.statd service would also have a fishy appearance about it. As such, you have to make sure the correct honeypots are running relevant services.

Another issue is of *where do you deploy* honeypot? A sophisticated attacker may come to know the location of honeypot in the network and if that happens, the entire network is under risk of attack. Next issue is how *many* honeypots will suffice? If the attack is high, large number of honeypots in the network would make it strong to mitigate the attack. They will provide higher quality of deception. On the other hand, if at some point of time, in the same network, attack becomes low and number of legitimate clients requesting services increase, then there is inefficient resource utilization. Same honeypots that are sitting idle waiting for attacks to happen could have been used to service client requests.

As if configuring and deploying was not enough, someone has to maintain the honeypot(s) once they go live. This can be more challenging then it sounds. Not only is honeypot technology rapidly developing and changing (requiring updates for your honeypots) but so are the networks. New systems are constantly being added (such as the latest Linux server) old systems removed or upgraded (such as that Novell NetWare server still running IPX), new applications introduced (Instant Messaging, P2P, or live video), while old services may be phased out (Gopher, Telnet). Your organization and networks are constantly changing. To stay current, honeypots have to adapt to the changes. Traditionally, this means someone manually updating or modifying the honeypots to better mirror the production environment [1]. That means time, money, and mistakes.

7.2 DYNAMIC HONEYPOTS

Dynamic honeypot is an autonomous honeypot capable of adapting in a dynamic and constantly changing network environment [1]. The dynamic honeypot approach integrates passive or active probing and virtual honeypots. This approach addresses the challenge of deploying and configuring virtual honeypots.

Dynamic honeypot, in near future would be a plug-n-play solution. You simply plug it in and the honeypot does all the work for you. It automatically determines how many honeypots to deploy, how to deploy them, and what they should look like to blend in with your environment. Even better, the deployed honeypots change and adapt to your environment. You add Linux to your network, you suddenly have Linux honeypots. You remove Novell from your network, your Novell honeypots magically disappear. You replace your Juniper routers with Cisco IOS, and so your honeypot routers change. The goal is an appliance, a solution you simply plug into your network, it learns the environment, deploys the proper number and configuration of honeypots, and adapts to any changes in your networks. Sound like magic? It shouldn't, the technology is there. We just have to put it together.

The first (and most critical) part of a dynamic honeypot is how it learns about the network, what systems your organization is using and how they are being used. With this knowledge, dynamic honeypot can intelligently map and respond to your environment. One possible approach is to actively probe the network, determine what systems are live, what type of systems they are, and the services they are using. Nmap is one such scanning tool with that capability. However, there are some drawbacks to such an active method. First, you are introducing more activity to your networks. Not only can you affect bandwidth or network activity, but with the scanning process you can cause services, even entire systems to shutdown. Second, its possible to miss a system, as it may be firewalled. Probes are sent to the system, but nothing is returned. Third, active scanning takes only a snapshot of a point in time; it cannot update you to real-time changes. To do that, you would constantly need to scan your environment. Not a very elegant approach. For our dynamic honeypot, one should take a passive approach, specifically passive fingerprinting [2] and mapping.

The concept of passive fingerprinting is not new. The idea is relatively simple, to map and identify systems on your network. However, you do not actively probe the systems, instead you passively capture network activity, analyze that activity, then determine the system's identity. The technology uses the same methods as active scanning [2]. Tools such as Nmap build a database of known operating systems and services. These tools then actively send packets to the target, packets which illicit a response. These responses (which are unique to most operating systems and services) are then compared to a database of known signatures to identify the operating system and services of the remote system. Passive fingerprinting takes the same approach, it has a database of known signatures for specific systems. However, the data is taken passively. Instead of actively probing remote systems, passive fingerprinting sniffs traffic from the network, then analyzes the packets from that network. It compares the packets against a database of signatures to identify the remote system (and potentially the services). Passive fingerprinting is not limited to TCP, other protocols can be used.

There are several advantages to using passive technologies. The first is that it's not intrusive. Instead of actively interacting with systems, you are passively gathering data. There is far less likelihood of you damaging or taking down a system or service. Second, even if systems are using host-based firewalls, passive fingerprinting will identify the system, if nothing else it will map a MAC address to an IP. Last, this method is continuous—as your networks changes, these changes can be captured in real time. This becomes critical for maintaining realistic honeypots over the long term. The disadvantage of passive mapping is it may not work well across routed networks; it's potentially more effective on your local LAN. Then again, this is true for active mapping also. In some cases, more then one dynamic honeypot would have to be physically deployed in your organization, depending on its size, number of networks, and configuration.

Dynamic honeypot [1] could leverage this concept of passive fingerprinting to learn networks. The honeypot could be deployed as an appliance or single box. This device is then physically connected to your network. Once connected, it spends the next 24 to 72 hours watching and learning your network. By passively analyzing all of the traffic it sees, it determine how many systems are on your networks, the operating system types, the services they offer, and potentially even which systems are communicating with whom and how often. This information is then used to learn and map your network. Once the honeypot learns the environment, it can begin deploying more honeypots. The advantage here is that the honeypots are crafted to mirror your environment. By looking and behaving the same way as your production environment, the honeypots seamlessly blend in, making them much more difficult for attackers to identify as honeypots, or to 'sniff them out'. However, this passive learning does not stop. Instead, it continuously monitors the network. Whenever changes are made, these changes are identified and the deployed honeypots adapt to the changes. If your organization is a typical Windows environment, you may begin deploying some Linux servers. Dynamic honeypot, using passive fingerprinting, can determine that Linux systems have been deployed. Honeypot would then deploy Linux honeypots, or update existing honeypots, based on the same Linux makeup and using similar services. The dynamic honeypot vastly reduces not only the work involved in configuring your honeypots, but also maintains them in a constantly changing environment.

A commercial example of such capabilities is Sourcefire's Real-Time Network Awareness. Designed to work with IDS sensors, RNA passively maps your network to coordinate IDS alerting with your network makeup. By utilizing tools such as these, honeypots can now learn and monitor their environments in real time.

The next problem to be solved is how do the honeypots get deployed? Passive fingerprinting offers a powerful tool, but how do we actually get it to populate our network with honeypots? Traditionally, this would require physically deploying a new computer for each IP address we wanted to monitor. However, this defeats the purpose of a dynamic honeypot if a person has to physically deploy multiple honeypots. We need a hands free, fire-and-forget solution. A far more simple and effective approach is not to deploy any physical honeypots. Instead, our honeypot appliance deploys hundreds, if not thousands, of virtual honeypots monitoring all of the unused IP space. All of these virtual honeypots are deployed and maintained by our single physical device. Because the virtual honeypots monitor unused IP space, we can be highly confident that any activity to or from those IPs is most likely malicious or unauthorized behavior. Based on previous passive mapping of the network, we can also determine how many honeypots we should deploy, the types, and where. For example, our passive mapping may have determined that on our Class C network, we have 100

Windows XP workstations, twenty Windows 2003 servers, five Linux servers, and two Cisco switches. Based on this, dynamic honeypot can create an equivalent ratio of honeypots. Perhaps ten Windows XP honeypots, two Windows 2003 servers, one Linux server, and one Cisco switch. Honeypots now not only match the type of production systems in use and their services, but the ratio of systems used. Not only that, but the virtual honeypots can also monitor the same IP space as the systems themselves. For example, perhaps honeypot learns that the Windows XP workstations are DHCP systems in the 192.168.1.100–192.168.1.250 range. Windows XP honeypots would reside in the same IP space, while the other honeypots are monitored their respective IP space.

Once again, this ability to dynamically create and deploy virtual honeypots already exists. The OpenSource honeypot Honeyd allows a user to deploy virtual honeypots throughout an organization. In addition, this honeypot can emulate over 500 operating systems, both at the IP stack and application level. As an OpenSource solution, its highly customizable, allowing it to adapt to almost any environment. By combining the capabilities of a solution like Honeyd, with the capabilities of a passive fingerprinting tool such as p0f, we come very close to our dynamic honeypot. We can have a hands free, fire-and-forget solution. You deploy your Honeyd honeypot by connecting it to your network. The passive fingerprinting tool p0f kicks in, passively monitoring and mapping your network. After a certain period of time, it learns what systems you have, what they are running, where they are located, and potentially how they are being used. Based on this data, your honeypot creates virtual honeypots that mirror the makeup of your network, and subtly blend in with your production systems. Attackers can no longer tell what a honeypot is and what part of your network is really. Once these honeypots are virtually deployed, p0f continues to monitor your networks. The virtual honeypots adapt in real time to any additions, changes, or removal of existing systems. A security administrator has to sit back and catch the bad guys.

7.3 DYNAMIC HONEYPOT DESIGN

One of the biggest challenges when deploying any type of a security system is in maintaining the functionality of the total system as the network topology or technology changes. This issue is especially critical for honeypots. It must be made sure that honeypots blend into the system and appear like any other entity on the network. The services should be chosen so as to mimic the real services that a certain OS is able to provide and keep up with new technologies while removing old and obsolete services.

What is sought is a dynamic honeypot that we can just plug in and leave it to operate without the need to constantly update it. It should be able to automatically identify unused IPs and deploy virtual honeypots on these IPs. Also any time a system is removed or added to the network; dynamic honeypots should be able to reconfigure themselves automatically.

7.3.1 Proposed Design Overview

To implement the dynamic Honeypot approach, the components listed below are needed [3]. The last three components represent main contribution to the design of a dynamic honeypot.

- An active probing tool, such as Nmap.
- A passive fingerprinting tool, such as P0f and Snort.
- A low interaction honeypot used to simulate networks, such as Honeyd.
- A collection of physical honeypots to receive redirected traffic. These high interaction honeypots are considered as a small but representative sample of the most used operating systems on the network.
- A Database, containing hosts' description, and log information.
- A Dynamic honeypot Engine, which interacts with the previous components to dynamically configure Honeyd and generate output (see Figure 7.1).
- An administrator's interface to configure dynamic honeypot server in real time and view reports.

Dynamic honeypot server starts by collecting information about the hosts available on the networks using active or passive approach. The administrator has the choice of selecting the best data gathering approach to be used based on the network architecture. If the network consists of computers connected through a hub where packet sniffing is feasible, the administrator would run dynamic honeypot server in passive mode avoiding the need for generating probing packets on the shared medium. On the other hand, if network consists of switched network where the hosts connected to layer two switch, passive fingerprinting would not be as reliable as active probing. After having the complete picture of the network including hosts' operating systems and services running, dynamic honeypot server estimates the personalities and services of the fake systems to be deployed and then issues the suitable configuration parameters to honeyd to deploy the systems on the network. The network will now have both real and fake systems running together. An intruder can be detected using the connections that are made to the fake systems which are non production systems and are not supposed to receive traffic from the network. Any attempt to interact with a fake system would be redirected to a physical honeypot permitting the hacker a higher level of interaction. The interaction with the physical honeypot is logged using Honeynet Sebek client and is sent to the dynamic honeypot server which uses sebek server to capture the logs. This approach allows central collection of logs so the administrator will be presented with summarized reports of interactivity on both the fake and physical honeypots. The dynamic honeypot server constantly analyzes the data available in the database and sends an alert (such as an SMS) to the administrator in case of a high level of risk and a need of immediate attention.

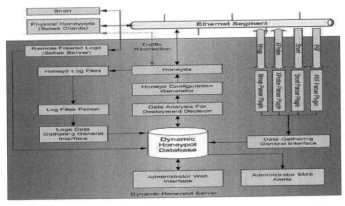

Figure 7.1 Dynamic Honeypot Server Design.

An important advantage of using the approach of dynamic honeypot lies in its ability to capture malicious attacks both on the small scale, i.e., from host perspective (physical honeypot logs) and the large scale, i.e., from the network perspective (virtual honeypot logs). For example, we can identify a worm from repeated scans on a certain port on multiple hosts. This combination of gathered logs permits the administrator to better analyze and classify attacks.

7.3.2 Active Probing

To implement the dynamic honeypot approach, one can actively probe the entire network and request responses from a targeted system in order to determine the OS and the services it uses. There are some drawbacks if active probing is extensively used. This excessive probing consumes additional bandwidth and may cause systems to shutdown.

A second issue arises when the targeted systems do not respond when a local firewall with a high security policy is in place. As mentioned before, active probing needs to keep requesting answers from the machines on the network in order to work in a real time fashion. Nmap, for example, is a tool equipped with a database of known operating systems and services. It actively sends packets to the target and illicit responses. These responses are then compared to known signatures in a database to identify the operating system and services of the remote system.

7.3.3 Passive Fingerprinting

An alternative approach is to use passive fingerprinting which is based on sniffing packets from the network to capture network activity, analyze it, and determine the fingerprinting of the system (OS and services). This will not bring the computer down or utilize extra bandwidth. Also the firewall problem will be alleviated since passive fingerprinting will identify the system and any changes in real time can still be captured. The concept of sniffing the network packets can introduce a problem when using routed or switched networks and therefore, it is more effective on a shared medium, such as a repeater hub.

POf is a passive fingerprinting program that compares SYN packets against a set of signatures stored in its database in its attempt to determine the OS type. POf is totally silent and does not impose traffic on the network; it uses TCP/IP fields to infer the type of OS used. This implementation of Dynamic Honeypot uses Snort to passively gather detailed information about the services running on the current network. The network traffic can be filtered using snort rules in order to gather only that information relevant to the current services running on the network.

7.3.4 Honeyd

This is a low-interaction honeypot[4]. Developed by Niels Provos, Honeyd is OpenSource and designed to run primarily on UNIX systems. Honeyd works on the concept of monitoring unused IP space. Anytime it sees a connection attempt to an unused IP, it intercepts the connection and then interacts with the attacker, pretending to be the victim. By default, Honeyd detects and logs any connection to any UDP or TCP port. When an attacker connects to the emulated service, not only does the honeypot detect and log the activity, but it captures all of the attacker's interaction with the emulated service. In the case of the emulated FTP server, we can potentially capture the attackers' login and password, the commands they issue, and perhaps even learn what they are looking for or their identity.

Honeyds can be used to simulate networks, and provides an IP stack simulated response of the OS they represent. They do not provide full application layer response such as is the case with a high interaction honeypot. They can redirect network activities that require a high level of interaction to a physical honeypot. For example, they can redirect DNS requests to a proper name server. If someone uses active fingerprinting measures to determine the OS type of the honeyd most honeyds respond with the IP stack of whatever OS they are configured to mimic. This is accomplished by spoofing the needed replies.

7.3.5 The Database

The database contains information about the real and virtual systems; it contains mainly 6 tables:

1. IPInformation Table: It contains the description of the systems having a certain IP address. The description of the system includes information about the operating system, the time the data was collected, in addition to a flag indicating whether the IP is used by a real system or a virtual system. This table also includes a code for the probing tool used, since some probing tools are more reliable than others.
2. Ports Table: Contains more detailed description of a certain IP address. The data includes a list of all ports that are open and the services. In addition there are entries for the probing tool used and the time the data was collected.
3. OperatingSystems Table: Contains a unique ID given to each operating system together with a description of that operating system.
4. ServicesScripts Table: Contains information about the honeyd script and the service it is used to simulate.
5. HoneydLog Table: In case a honeyd is accessed, the log data is filled into this table. The table contains information about the packets received by the honeyds, including the source and destination addresses and their respective ports. In addition there are entries for information about time and possibly some information about the intruder's operating system or his scanning tool.
6. Sebek Logs Table: This is used as part of Sebek Server to capture interactivity with physical honeypots.

7.3.6 Dynamic Honeypot Engine

This module performs 4 main functions:

1. Gather the needed information required for honeyd deployment and perform necessary calculations: In order to be able to deploy the right combination of low interaction honeypots, with the right characteristics, we need to gather

information about the real systems. This is done using the IPInformation table. The calculations are made such that the virtual systems reflect the ratio of the real systems' operating systems. Reflecting the ratio means that the number of used IPs running a certain OS to the total number of used IPs must match the number of deployed virtual honeypots simulating the same OS to the total number of the deployed virtual honeypots. For example if we have 5 used IPs with 2 Windows NT and 3 Red Hat Linux, and we have 10 free IPs these should be filled with 4 virtual Windows NT and 6 virtual Red Hat Linux.

2. Issue configuration commands for deployment of honeyds: Now that the engine knows what low interaction honeypots should be deployed, configuration commands are sent to honeyd to deploy the virtual systems.

3. Collect logs and generate output reports: The administrator is alerted if there is any attack. The data in the HoneydLog should also be reported, this allows the administrator to view how often honeyds are accessed and the corresponding details of attack.

4. Delete Expired database entries: This functionality basically checks for the time field for entries in the database and compares them with the default expiration time or to the custom period set by the administrator.

7.4 DYNAMIC HONEYPOT CONSTRUCTION

The Honeyd Configuration Manager system[5] was implemented in Perl using the Linux OS, as shown in Figure 7.2. The system actively scans the network using Nmap for OS detection and to determine which TCP/UDP ports which are open. While it does consume some bandwidth, active scanning quickly provides sufficient data from which a honeypot or Honeynet configuration file can be constructed. The alternative approach of using passive scanning, such as that performed by p0f, only provides information about ports that are communicating, and as a result gathers information about the network and active hosts more slowly. Passive scanning also has limitations in the network topology which it can monitor,

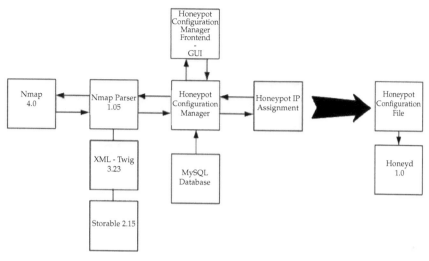

Figure 7.2 Honeyd Configuration manager components.

including issues related to Layer 2 switches and routers. Passive scanning requires the detection of packets on the network to determine the network configuration by looking at the packet headers. It is necessary that the scanner intercept all packets for a more complete picture. This is made more challenging with the use of Layer 2 switches and routers which, for security and efficiency, isolate the different devices. Routers can also confuse the passive scanner by modifying the packet headers. The default Nmap scan used by the Honeyd Configuration Manager is a SYN stealth scan, RPC scan, UDP scan, and OS detection, as shown in the following Nmap command:

nmap -sS -sU -sR –O <port range>

Other scan options are available in Nmap, and can be selected for the Honeyd Configuration Manager using either the GUI or the CLI. The Nmap results are analyzed real-time by the Honeyd Configuration Manager to determine the resulting honeynet configuration, which is stored as a Honeyd configuration file. As the Honeyd file is created the program uses the IP assignment method selected by the administrator to determine the network configuration of the honeypots. These IP assignments allow the administrator to create replicas of the scanned machines and networks in a manner most appropriate to the task, such as research into attack behavior or the deployment of honeypots in a production environment. The four options for IP address assignment are as follows:

1. Configure the honeypots to use the same IP addresses as the real systems. For example, if a real system is found with the IP address 192.168.0.7, a system will be added to the Honeyd configuration file with the same address. This option should not be used to create a honeynet which will exist on the same network as the production systems as it will result in IP address conflicts. However, it can be a useful starting point for creating a Honeyd configuration file prior to manual modification, or if the honeynet will be used when the production systems are isolated from the network or powered down (overnight, for example in some circumstances).

2. Modify the network component of the IP address, while preserving the host component. For example, if a host with IP address of 192.168.0.12 is found while scanning a class C network, the network component can be mapped to another subnet, such as 172.16.0.0/24, resulting in a system in the Honeyd configuration file with IP address 172.16.0.12. This would be useful when creating a Honeynet.

3. Configuration for use outside of the production network, such as for testing in a laboratory or other controlled environment.

4. Use a designated IP address range in which to place the resulting honeypots. For example, if the user designates 172.16.0.20-40 as the honeypot IP address range, then the first production host found, such as 192.168.0.25, would result in a system in the Honeyd configuration file with the IP address 172.16.0.20. Honeypot representations of additional production hosts identified in the scan would continue to receive IP addresses in the assigned range until it was completely populated.

5. Interweave the honeypot systems into the production network where possible, although this requires that the assigned IP addresses on a given production network can be identified. If the scanning system is on the same subnet as the systems being scanned then unassigned IP address can be identified by whether or not there is a response to an ARP packet. In the event that the scanning system is not on the same subnet as the systems being scanned other approaches can be used

to identify unassigned IP address, including pings or reverse DNS lookups, although these are likely to provide inaccurate results so the resulting Honeyd configuration file should be used with caution. Regardless of the method used to identify the unassigned IP addresses, if new systems are added to the production network it is possible that an IP address conflict may result, depending on the method used to assign address to new systems.

This approach is most useful when building a honeynet configuration for use in conjunction with existing production systems, such as in a commercial environment. Once the production systems, with their operating systems and open ports, have been identified, the Honeyd configuration file is created and used to start, or restart, the honeynet. The reconfiguration of the honeynet usingthis method can be invoked manually by the Honeynet administrator, or configured to execute automatically on a periodic schedule or in response to an identified asynchronous event, such as the assignment of a new DHCP address, which may suggest that a new system has been added to the environment. In addition to providing a configuration which emulates the production hosts and open ports, it is possible to add scripts to Honeyd which can provide a more in-depth emulation of a given service. For this research effort, a small MySQL database has been used to store the details about the available scripts, including which services they emulate, and for what operating systems they are appropriate. Using this approach, when an open port is found by the Honeyd Configuration Manager is possible to quickly query the database for an appropriate script. For each script, the database includes information about the operating system(s) and port(s) for which it is appropriate, to ensure, for example, that a script which emulates the Microsoft IIS 5.0 Web Server does notappear on a Linux based honeypot. If a suitable script is found, it can then be automatically added to the Honeyd configuration to provide a more in-depth emulation (i.e., higher interaction level) on the particular port. The MySQL database used for testing contained six emulation scripts, but this database-based framework allows for addition scripts to be utilized by simply adding new records to the database. A basic example of a Honeyd configuration file created by the Honeyd Configuration Manager is shown in Figure 7.2. This honeynet consists of a single honeypot, which is a Microsoft Windows based system, with open several ports, and a telnet emulation script listening on port 23. Typical Honeyd configuration files for production networks are substantially larger, and commonly feature honeypots using several different operating systems, openports, and emulated services.

7.4.1 Graphic User Interface

A Graphical User Interface (GUI) was also created as a front-end for the command line based Honeyd Configuration Manager. This allows the user to easily control the behavior of the Honeyd Configuration Manager, including the IP assignment mechanism, Nmap scan parameters, and whether service emulation scripts should automatically be used. In addition, the GUI allows the user to monitor the results of the scanning process, and to select whether the tool should merely create a Honeyd configuration file, or should also start a Honeyd instance at the completion of the configuration process. The GUI was implemented using Glade User Interface Builder for GTK+ and GNOME. Figure 7.3 shows an example of the GUI in use.

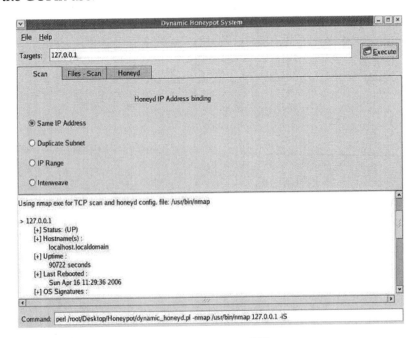

Figure 7.3 The HoneyD configuration manager GUI.

7.5 BENEFITS

The use of dynamically configured honeypots or Honeynets provides several advantages over static configurations:

1. A system administrator or researcher can quickly build a honeypot without the need to have in-depth knowledge of the honeypot configuration mechanism, which in the case of Honeyd takes the form of a text configuration file. The dynamically constructed configuration can then either be used to deploy a honeypot, or as a basis for refinement by the honeynet administrator.

2. The honeynet administrator is not required to know the details of the network topology or installed systems in the network to be mimicked, as this information can be automatically ascertained.

3. The honeynet configuration can be easily reconfigured to reflect the current network topology. This reconfiguration can either be performed on a predefined schedule, asynchronously in response to some event such as the assignment of a new IP address by a DHCP server, or even in near-real time if desired.

7.6 SUMMARY

Dynamic honeypots can radically revolutionize the deployment and maintenance of honeypots. By learning and monitoring networks in real time, they become a fire-and-forget solution. Not only do they become cost-effective to deploy and maintain, but they have better integration into your network. The design includes a dynamic honeypot engine that integrates data collected from passive fingerprinting tools such as p0f and active probing tools such as Nmap to dynamically configure Honeyds.

In the field of security, honeypot have emerged as successful tools and technology to be used in conjunction with IDS and firewalls. The flexibility of honeypots allows them to be used for adding value to the security of an organization in production network or just for research purposes. Research honeypots contribute little to the direct security of an organization. They can be deployed outside the network, in DMZ or inside the network according to the goals and requirements and serve different purposes. Honeypots are an early detection mechanism, however, they add risk of attack in the network they are deployed.

EXERCISES

1. Outline the requirements of having dynamic honeypots in the network. What are the major issues in making the honeypots dynamic?
2. Explain the design of dynamic honeypot.
3. What is active probing? How is it different from passive fingerprinting?
4. Describe the various components of Honeyd configuration manager.
5. Explain the database tables used in Dynamic Honeypot Server Design.

REFERENCES

[1] R. Budiarto, A. Samsudin, C.W. Heong, and S. Noori, "Honeypots: why we need a dynamics honeypots?," in *International Conference on Information and Communication Technologies*, 2004, pp. 565–566.

[2] G. Conti and K. Abdullah, "Passive visual fingerprinting of network attack tools," in *Proceedings of the 2004 ACM workshop on Visualization and data mining for computer security* Washington DC, USA 2004, pp. 45–54.

[3] I. Kuwatly, M. Sraj, Z. Al Masri, and H. Artail, "A dynamic honeypot design for intrusion detection," in *Pervasive Services, 2004. ICPS 2004. IEEE/ACS International Conference on* 2004, pp. 95–104.

[4] N. Provos, "Honeyd-a virtual honeypot daemon," 2003.

[5] C. Hecker, K.L. Nance, and B. Hay, "Dynamic honeypot construction," 2006.

8

Wireless Honeypots

This chapter will introduce honeypots as a countermeasure for attacks in wireless networks. It starts with an introduction to wireless networks and 802.11 attacks paradigm. Next it gives an overview of wireless honeypots history and then moves on to addressing related goals. It delves further into wireless honeypot architectures and implementation using wired tools. Then it discusses the wireless honeypot design. It then explains the theoretical aspects and design possibilities, before looking at two easy technical examples. Lastly, some of the limitations of wireless architectures are discussed.

Wireless technologies have redefined the contours of our world, a world without wires where information is available from anywhere at any time. The huge expansion of wireless technologies, and specifically 802.11 wireless data networks, in the last few years has provided a new battle field for information access. End users and corporations are heavily interested in taking advantage of the flexibility, mobility and freedom offered by wireless technologies to access and share information. The human dependency on those technologies has increased to the point where one can find wireless devices almost everywhere, from network devices to laptops, cameras, and so on. These allow connecting to data networks, like the Internet.

Along with this freedom, though, come security issues that must be thoroughly understood and addressed. Though wireless devices support standard security options and protocols useful to thwart common attacks (ciphering, authentication, etc.), many kinds of attacks are still possible but are dependent on the real level of security present and the skill of the attacker. Due to the exponential usage of wireless equipment and technologies today, it is required to get an in-depth knowledge about the real exploitation vectors currently used to compromise wireless networks. Trying to fill this knowledge gap, the main goal of Wireless honeypots is to analyze the state of real life wireless hacking and thereby making the networks more secure.

Wireless honeypots are a way to try and stay ahead of the hacker community, helping to determine the latest tools and techniques being used by the hacker community.

8.1 INTRODUCTION TO WIRELESS LOCAL AREA NETWORKS

A Wireless Local Area Network (WLAN) links two or more devices using some wireless distribution method (typically spread-spectrum or OFDM radio), and usually provides a connection through an AP to the wider internet. This gives users the mobility to move around within a local coverage area and still be connected to the network as shown in Figure 8.1.

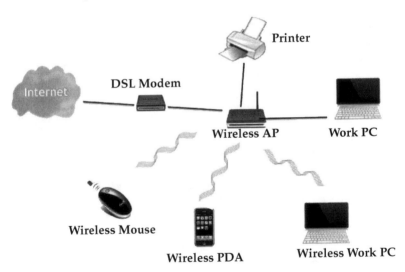

Figure 8.1 Wireless Local Area Network (WLAN).

IEEE 802.11 is a set of standards carrying out wireless local area network (WLAN) computer communication in the 2.4, 3.6 and 5 GHz frequency bands [1]. They are created and maintained by the IEEE LAN/MAN Standards Committee (IEEE 802). The base current version of the standard is IEEE 802.11-2007. The original version of the standard IEEE 802.11 was released in 1997 and clarified in 1999, and is now obsolete. Legacy of 802.11 with direct-sequence spread spectrum was rapidly supplanted and popularized by 802.11b.

802.11b

IEEE expanded on the original 802.11 standard in July 1999, creating the 802.11b specification. 802.11b supports bandwidth up to 11 Mbps, comparable to traditional Ethernet.

802.11b uses the same unregulated radio signaling frequency (2.4 GHz) as the original 802.11 standard. Vendors often prefer using these frequencies to lower their production costs. Being unregulated, 802.11b gear can incur interference from microwave ovens, cordless phones, and other appliances using the same 2.4 GHz range. However, by installing 802.11b gear a reasonable distance from other appliances, interference can easily be avoided. Some of the advantages of 802.11b are its low cost and good signal range however home appliances may interfere on the unregulated frequency band.

802.11a

While 802.11b was in development, IEEE created a second extension to the original 802.11 standard called 802.11a. 802.11b gained popularity much faster than did 802.11a, some folks believe that 802.11a was created after 802.11b. In fact, 802.11a was created at the same time. Due to its higher cost, 802.11a is usually found on business networks whereas 802.11b better serves the home market.

802.11a supports bandwidth up to 54 Mbps and signals in a regulated frequency spectrum around 5 GHz. This higher frequency compared to 802.11b shortens the range of 802.11a networks. The higher frequency also means 802.11a signals have more difficulty penetrating walls and other obstructions.

Because 802.11a and 802.11b utilize different frequencies, the two technologies are incompatible with each other. Some vendors offer hybrid 802.11a/b network gear, but these products merely implement the two standards side by side (each connected devices must use one or the other). Its high speed and regulated frequencies prevent signal interference from other devices however the cost of the equipment is very high.

802.11g

In 2002 and 2003, WLAN products supporting a newer standard called 802.11g emerged in the market. 802.11g attempts to combine the best of both 802.11a and 802.11b. 802.11g supports bandwidth up to 54 Mbps, and it uses the 2.4 Ghz frequency for greater range [2]. 802.11g is backwards compatible with 802.11b, meaning that 802.11g APs will work with 802.11b wireless network adapters and vice versa. Its speed is high; in addition to that the signal range is good and not easily obstructed.

802.11n

The latest IEEE standard in the WLAN category is 802.11n. It was designed to improve on 802.11g in the amount of bandwidth supported by utilizing multiple wireless signals and antennas called Multiple Input and Multiple Output technology (MIMO) instead of one. The IEEE has approved the amendment and it was published in October 2009. 802.11n connections supports data rates of over 100 Mbps. 802.11n also offers somewhat better range over earlier WLAN standards due to its increased signal intensity. 802.11n equipment is backward compatible with 802.11g gear. It has a better signal range and is more resistant to signal interference from outside sources however the use of multiple signals may greatly interfere with nearby 802.11b/g based networks.

Table 8.1 802.11 Standards.

802.11 Proto-col	Release	Freq (GHz)	Bandwidth (Mhz)	Data rate per Stream (Mbit/s)	Modu-lation	Approx Indoor Range(m)	Approx Outdoor Range(m)
-	Jun 1997	2.4	20	1, 2	DSSS	20	100
A	Sept 1999	5	20	6, 9, 12, 18, 36, 48, 54	OFDM	35	120
						—	5,000
B	Sept 1999	2.4	20	1, 2, 5.5, 11	DSSS	38	140
G	Jun 2003	2.4	20	1, 2, 6, 9, 12, 18, 24, 36, 48, 54	OFDM, DSSS	38	140
N	Oct 2009	2.4/5	20	7.2, 14.4, 21.7, 28.9 43.3, 57.8, 65, 72.2	OFDM	70	250
			40	15, 30, 45, 60, 90, 120, 135, 150		70	250

8.2 BASIC WIRELESS CONCEPTS

8.2.1 Stations and APs

A wireless network interface card (adapter) is a device, called a station, providing the network physical layer over a radio link to another station. An Access Point (AP) is a station that provides frame distribution service to stations associated with it. The AP itself is typically connected by wire to a LAN.

The station and AP each contain a network interface that has a Media Access Control (MAC) address, just as wired network cards do. This address is a world-wide-unique 48-bit number, assigned to it at the time of manufacture. The 48-bit address is often represented as a string of six octets separated by colons (e.g., 00:02:2D:17:B9:E8) or hyphens (e.g., 00-02-2D-17-B9-E8). While the MAC address as assigned by the manufacturer is printed on the device, the address can be changed in software.

Each AP has a 0 to 32 byte long Service Set Identifier (SSID) that is also commonly called a network name. The SSID is used to segment the airwaves for usage. If two wireless networks are physically close, the SSIDs label the respective networks, and allow the components of one network to ignore those of the other. SSIDs can also be mapped to virtual LANs; thus, some APs support multiple SSIDs. Unlike fully qualified host names (e.g., gamma. cs.wright.edu), SSIDs are not registered, and it is possible that two unrelated networks use the same SSID.

8.2.2 Infrastructure and Ad Hoc Modes

A wireless network operates in one of the two modes, i.e., infrastructure or ad-hoc mode. In the ad hoc mode, each station is a peer to other stations and communicates directly with other stations within the network as shown in Figure 8.2. No AP is involved. All stations can send Beacon and Probe frames. The ad hoc mode stations form an Independent Basic Service Set (IBSS).

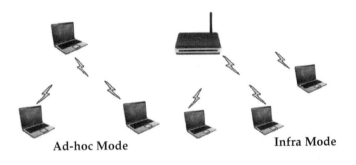

Ad-hoc Mode   Infra Mode

Figure 8.2 Infrastructure and ad-hoc modes.

A station in the infrastructure mode communicates only with an AP. Basic Service Set (BSS) is a set of stations that are logically associated with each other and controlled by a single AP. Together they operate as a fully connected wireless network. The BSSID is a 48-bit number of the same format as a MAC address. This field uniquely identifies each BSS. The value of this field is the MAC address of the AP.

8.2.3 Frames

Both the station and AP radiate and gather 802.11 frames as needed. The format of frames is illustrated below in Figure 8.3. Most of the frames contain IP packets. The other frames are for the management and control of the wireless connection.

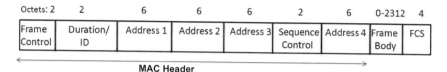

Figure 8.3 IEEE 802.11 Frame.

There are three classes of frames. The management frames establish and maintain communications. These are of Association request, Association response, Reassociation request, Reassociation response, Probe request, Probe response, Beacon, Announcement traffic indication message, Disassociation, Authentication, Deauthentication types. The SSID is part of several of the management frames. Management messages are always sent in the clear, even when link encryption (WEP or WPA) is used, so the SSID is visible to anyone who can intercept these frames.

The control frames help in the delivery of data.

The data frames encapsulate the OSI Network Layer packets. These contain the source and destination MAC address, the BSSID, and the TCP/IP datagram. The payload part of the datagram is WEP-encrypted [3].

8.2.4 Authentication

Authentication is the process of proving the identity of a station to another station or AP. In the open system authentication, all stations are authenticated without any checking. A station A sends an Authentication management frame that contains the identity of A, to station B. Station B replies with a frame that indicates recognition, addressed to A. In the closed network architecture, the stations must know the SSID of the AP in order to connect to the AP. The shared key authentication uses a standard challenge and response along with a shared secret key [4].

8.2.5 Association

Data can be exchanged between the station and AP only after a station is associated with an AP in the infrastructure mode or with another station in the ad hoc mode. All the APs transmit Beacon frames a few times each second that contain the SSID, time, capabilities, supported rates, and other information. Stations can choose to associate with an AP based on the signal strength, etc., of each AP. Stations can have a null SSID that is considered to match all SSIDs.

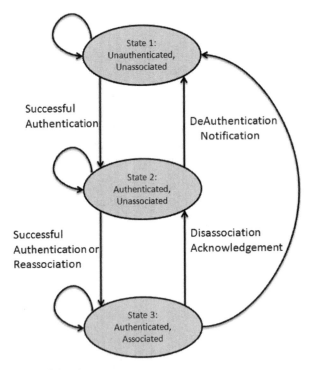

Figure 8.4 States and Services.

The association and authentication is a three-step process as shown in Figure 8.4. A station that is currently unauthenticated and unassociated listens for Beacon frames. The station selects a BSS to join. The station and the AP mutually authenticate themselves by exchanging Authentication management frames. The client is now authenticated, but unassociated. In the second step, the station sends an Association Request frame, to which the AP responds with an Association Response frame that includes an Association ID to the station. The station is now authenticated and associated.

A station can be authenticated with several APs at the same time, but associated with at most one AP at any time. Association implies authentication. There is no state where a station is associated but not authenticated [4].

8.3 802.11 SECURITY

Due to the RF signal nature of the wireless network, it is very difficult to control which computers or devices are receiving the wireless network signal. Therefore, the wireless relies on software link-level protection, specifically implementing cryptography to protect from eavesdropping and other network attacks. The original 802.11 standard only offers WEP to secure the wireless network.

8.3.1 Access Control List

The access control list is the simplest security measure one can find in a wireless network. The protection offered by this mechanism mainly consists of filtering out unknown users and requires a list of authorized client's MAC addresses to be loaded in the AP. Only those registered MAC addresses will be able to communicate with the AP, and will drop any communication that come from other unregistered MAC addresses.

As the name of this mechanism implies, it just offers access control to the network behind the AP, nor any other protection are offered, meaning that it will not protect each wireless client nor the wireless network traffic confidentiality or integrity. Therefore, any wireless network protected with only this mechanism is very vulnerable and open to any network attack.

8.3.2 WEP (Wired Equivalent Privacy)

Initially, WEP (Wired Equivalent Privacy) was the only link-level security option defined in the 802.11 standard. Its main purpose was the protection of the confidentiality and integrity of the wireless network traffic. WEP was designed to provide comparable confidentiality to a traditional wired network.

To meet the security proposed, WEP uses encryption to protect the data. WEP uses the RC4 stream cipher with a 64 or 128 bits key to provide data packet encryption. In addition, WEP can be used as access control method, because once WEP is active, the AP will just establish communication with nodes that have the same shared secret key, rejecting the others.

In a WEP-protected wireless network, every member of this wireless network, the wireless clients, must share the same secret key with the AP. This secret key can be a password or a character sequence generated by a wireless configuration program using a passphrase. It is not relevant how the secret key is introduce, especially true when using cards from several manufacturers, the most important thing is that every member of the wireless network must have the same WEP secret key.

WEP uses a 64 or 128 bits key to encrypt the data, but actually, the effective key is smaller, because part of the WEP key is transmitted in clear text along with the data packet. The WEP key, the key used to encrypt the data packet, is a concatenation of two values, a dynamic value called Initialization Vector (IV) and the static part of the key, the shared secret key. The Initialization Vector (IV) is a dynamic 24-bit value chosen randomly by the transmitter wireless network interface to give the WEP key liveness, giving more than 16 million possible keys. Liveness is required in WEP key because each message must be encrypted with a different key. The length of the shared secret key depends on the WEP key size chosen. When a 64 bit key is used, the shared secret key is 40 bits long (5 ASCII character) and for a 128 bit key, the shared secret key is 104 bits long (13 ASCII character). To synchronize the WEP key, the transmitter with the receiver, the IV value is transmitted in clear text along with data packet, therefore revealing 24 bits of the 64 or 128 bits key. In a WEP-protected wireless network, a data packet that gets into the link-level is encrypted before sending it off the air. The data packet encryption is performed as shown in Figure 8.5.

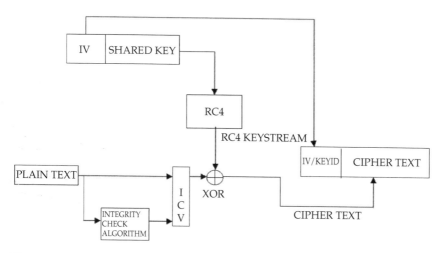

Figure 8.5 Wired Equivalent Privacy.

First, the wireless network interface randomly chooses an IV value and concatenated with the shared secret key to form the WEP Key (IV + secret key). The 802.11 standard does not define how the wireless network interface must be chosen. Therefore, the way the wireless network interface chooses the IV value depends on the manufacturer.

When the WEP Key is ready, this key is passed to the RC4 stream cipher to produce a pseudo-random string with the data packet length. The actual encryption occurs when the wireless network interface performs a logical XOR operation between the data packet with the pseudo-random string. To complete the WEP-protected data packet, the link-level headers, IV value and the encrypted data packet are packed together and then transmitted to the recipient. The steps for encrypting messages are also the same steps for decrypting a WEP-protected data packet. When the recipient decrypts a WEP-protected data packet, it first reads the IV value and then follows the same steps of the encryption process. A particular WEP key (the same IV and shared secret key) will always produce the same pseudo-random string, making it possible to regenerate the same pseudo-random string.

Finally, the data is decrypted by performing a second logical XOR operation between the encrypted data packet with pseudo-random string, cancelling the effect of the first logical XOR operation.

When the WEP was designed, it was considered sufficiently secure, until a research found out a weakness that was inherent in the WEP mechanism. From that time ever, several tools have been developed for cracking successfully the WEP shared secret key, turning it insecure [5].

WEP Key Recovery

The WEP protocol contains a critical cryptographic weakness that allows an attacker to possibly recover the shared secret key. This weakness consists of exploiting the fact that certain IV value produces weak WEP keys. When a weak WEP key is used to encrypt, the first bytes of the pseudo-random string may contain some correlation with the WEP key. Each weak key may leak one byte of the shared secret key with a 5% certainty probability. Because the certainty level is not high, the attacker must capture large amounts of raw WEP-protected data packets, approximately from 5 Million to 10 Million packets, to elevate the probabilities for successfully crack the shared secret key. This weakness was published by Fluhrer, Mantin,and Shamir, in "Weaknesses in the Key Scheduling Algorithm in RC4 [6]".

IV Collision

Another way for an attacker to break into a wireless network without knowing the shared secret key is also by capturing passively a large amount of data packets, but this time looking for IV collision. An IV collision occurs when two or more data packets are encrypted with the same IV value, therefore the same WEP key.

When an IV collision is detected, the attacker can perform a logic XOR with the two encrypted data packet to take off the encryption. The result is the XOR of the two data packets. With enough time, and use of analytic and statistics methods the attacker could be able to recover the contents of the two packets. The data packet contents recovery effort and time required decreases when more packets encrypted with exactly the same WEP key are captured and used for the recovery. Note that if this attack is successfully accomplished, the pseudo-random string recovery for a particular

WEP key is straightforward. An attacker could then keep all the pseudo-random strings in a record, using it for decryption, packets forging, and access to the network [5].

Some wireless interface card vendors, aware of the WEP key vulnerability, offer an update for their wireless cards to address the weak key vulnerability by avoid using weak IVs for encryption. Avoiding some IVs reduces the number of WEP keys making the IV collision weakness worse.

8.3.3 WPA (Wi-Fi Protected Access)

WPA is a solution released by the Wi Fi Alliance while a definitive security protocol is standardized. This security protocol is based on 802.11i, the next 802.11 wireless network security protocol standard. WPA consists of three main components: TKIP, 802.1x, and MIC. Each component was designed and implemented to address specific 802.11 weakness. Important security improvements were implemented, such as key hierarchy that protects and practically nullifies the exposure as WPA main key from attacks and implementing 802.1x protocol for access control to the network. Using key hierarchy means that WPA does not directly use the main key to encrypt, instead the main key (Pairwise Master Key) is used to generate other temporal keys such as session keys, group keys, etc.; and recursively the session key is used to generate the per-packet encryption key [7]. The IV is also expanded from 24 bits to 48 bits long and assigning it another role as a sequence counter for avoiding replay attacks. Improvement in packet integrity protection is made by implementing a specially designed cryptographically protected hashing function instead of using the CRC32 linear function.

8.3.4 802.11i

As mentioned earlier, WPA is a subset of the new security standard 802.11i (a.k.a. WPA2), meaning that 802.11i includes all WPA capabilities features and more security features. The main difference between 802.11i and WPA is the ability of 802.11i to use state-of-the-art AES (Advanced Encryption Standard) to encrypt the data packets. The AES algorithm is the encryption standard used by U.S government agency [8]. The downside of using AES encryption is that WEP only capable wireless network interface cannot be software upgraded to support it. AES. A wireless network that wants to use the 802.11i standard with full capabilities may require the replacement of the wireless network devices.

8.4 WIRELESS ATTACKS: FROM A BLACKHAT'S PERSPECTIVE

In general, attacks on wireless networks fall into four basic categories: passive attacks, active attacks, man-in-the middle attacks, and jamming attacks. Let's review what these attacks mean in a wireless network.

8.4.1 Passive Attacks on Wireless Networks

A passive attack occurs when someone listens to or eavesdrops on network traffic. Armed with a wireless network adaptor that supports promiscuous mode, the eavesdropper can capture network traffic for analysis using easily available tools, such as Network Monitor in Microsoft products, or TCPdump in Linux-based products, or AirSnort. A passive attack on a wireless network may not be malicious in nature. In fact, many in the wardriving community claim their wardriving activities are benign or "educational" in nature. It is worth noting that wardriving, looking for and detecting wireless traffic, is probably not illegal, even though propagandistic claims to the contrary are often made. Wireless communication takes place on unlicensed public frequencies. This makes protecting a wireless network from passive attacks more difficult.

Most of these wardriving enthusiasts use a popular freeware program, called Netstumbler. The Netstumbler program works primarily with wireless network adaptors that use the Hermes chipset because of its ability to detect multiple APs that are within range and WEP, among other features. The most common card that uses the Hermes chipset for use with Netstumbler is the ORiNOCO gold card. Another advantage of the ORiNOCO card is that it supports the addition of an external antenna, which can greatly extend the range of a wireless network to many orders of magnitude, depending on the antenna. A disadvantage of the Hermes chipset is that it doesn't support promiscuous mode, so it cannot be used to sniff network traffic. For that purpose, you need a wireless network adaptor that supports the PRISM2 chipset. The majority of wireless network adaptors targeted for the consumer market use this chipset, for example, the Linksys WPC network adaptors. Sophisticated wardrivers will arm themselves with both types of cards, one for discovering wireless networks and another for capturing the traffic.

Netstumbler is a sophisticated and feature-rich product that is excellent for performing wireless site surveys, whether for legitimate purposes or not. Not only can it provide detailed information on the wireless networks it detects, it can be used in combination with a GPS to provide exact details on the latitude and longitude of the detected wireless networks. Figure below shows the interface of a typical Netstumbler session [9].

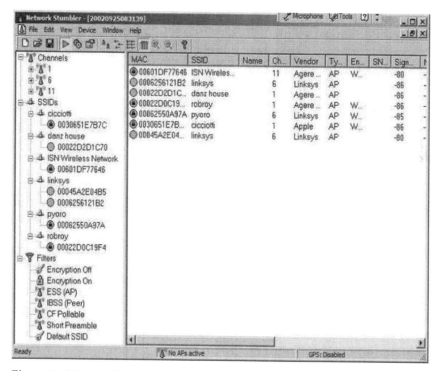

Figure 8.6 Netstumbler Screenshot.

Using a tool such as Netstumbler is only a preliminary step for the attacker. After discovering the SSID and other information, the attacker can connect to the wireless network to sniff and capture network traffic. This network traffic can reveal a lot of information about the network and the company that uses it. For example, looking at the network traffic, the attacker can determine what DNS servers are being used, the default home pages configured on browsers, network names, logon traffic, and so on. The attacker can use this information to determine if the network is of sufficient interest to proceed further with other attacks. Furthermore, if the network is using WEP, the attacker can, given enough time, capture a sufficient amount of traffic to crack the encryption.

8.4.2 Active Attacks on Wireless Networks

Once an attacker has gained sufficient information from the passive attack, the hacker can then launch an active attack against the network. There are a potentially large number of active attacks that a hacker can launch against a wireless network. For the most part, these attacks are identical to the kinds of active attacks that are encountered on wired networks. These include, but are not limited to, unauthorized access, spoofing, and Denial of Service (DoS) and Flooding attacks, as well as the introduction of Malware and the theft of devices. With the rise in popularity of wireless networks, new variations of traditional attacks specific to wireless networks have emerged along with specific terms to describe them, such as "drive-by spamming" in which a spammer sends out tens or hundreds of thousands of spam messages using a compromised wireless network.

Due to the nature of wireless networks and the weaknesses of WEP, unauthorized access and spoofing are the most common threats to a wireless networks. Spoofing occurs when an attacker is able to use an unauthorized station to impersonate an authorized station on a wireless network. A common way to protect a wireless network against unauthorized access is to use MAC filtering to allow only clients that possess valid MAC addresses access to the wireless network. The list of allowable MAC addresses can be configured on the AP, or it may be configured on a RADIUS server that the AP communicates with. However, regardless of the technique used to implement MAC filtering, it is a relatively easy matter to change the MAC address of a wireless device through software to impersonate a valid station. In Windows, this is accomplished with a simple edit of the registry, in UNIX through a root shell command. MAC addresses are sent in the clear on wireless networks, so it is also a relatively easy matter to discover authorized addresses.

WEP can be implemented to provide more protection against authentication spoofing through the use of Shared Key authentication. However, Shared Key authentication creates an additional vulnerability. Because Shared Key authentication makes visible both a plaintext challenge and the resulting ciphertext version of it, it is possible to use this information to spoof authentication to a closed network.

Once the attacker has authenticated and associated with the wireless network, he or she can then run port scans, use special tools to dump user lists and passwords, impersonate users, connect to shares, and, in general, create havoc on the network through DoS and Flooding attacks. These DoS attacks can be traditional in nature, such as a ping flood, SYN, fragment, or Distrbuted DoS (DDoS) attacks, or they can be specific to wireless networks through the placement and use of Rogue APs to prevent wireless traffic from being forwarded properly.

8.4.3 Man-in-the-Middle Attacks on Wireless Networks

Placing a rogue AP within range of wireless stations is wireless-specific variation of a man-in-the-middle attack. If the attacker knows the SSID in use by the network (which as we have seen is easily discoverable) and the rogue AP has enough strength, wireless users will have no way of knowing that they are connecting to an unauthorized AP as shown in Figure 8.7.

Using a rogue AP, an attacker can gain valuable information about the wireless network, such as authentication requests, the secret key that may be in use, and so on. Often, the attacker will set up a laptop with two wireless adaptors, in which one card is used by the rogue AP and the other is used to forward requests through a wireless bridge to the legitimate AP. With a sufficiently strong antenna, the rogue AP does not have to be located in close proximity to the legitimate AP [9]. So, for example, the attacker can run the rogue AP from a car or van parked some distance away from the building. However, it is also common to set up hidden rogue APs (under desks, in closets, etc.) close to and within the same physical area as the legitimate AP. Because of their undetectable nature, the only defense against rogue APs is vigilance through frequent site surveys using tools such as Netstumbler and AiroPeek, and physical security.

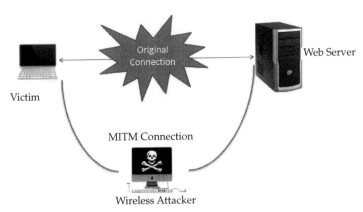

Figure 8.7 Man in the Middle Attack.

8.4.4 JAMMING ATTACKS ON WIRELESS NETWORKS

Jamming is a special kind of DoS attack specific to wireless networks.

Jamming occurs when spurious RF frequencies interfere with the operation of the wireless network. In some cases, the jamming is not malicious and is caused by the presence of other devices, such as cordless phones, that operate in the same frequency as the wireless network. In a case like this, the administrator must devise and implement policies regarding the use of these devices, such as banning the use of Bluetooth devices, or choosing wireless hardware that uses different frequencies. Intentional and malicious jamming occurs when an attacker analyzes the spectrum being used by wireless networks and then transmits a powerful signal to interfere with communication on the discovered frequencies as shown in Figure 8.8.

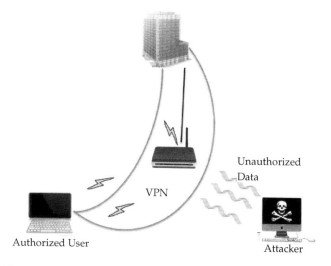

Figure 8.8 Jamming Attack.

Fortunately, this kind of attack is not very common because of the expense of acquiring hardware capable of launching jamming attacks. Plus, jamming a network represents a kind of pyrrhic victory for the attacker—a lot of time and effort expending merely to disable communications for a while.

8.4.5 Some other Attacks

Client-Client Attacks: Clients exist frequently on both wired and wireless networks. A client may be anything from a NAS (Network Attached Storage) device to a printer or a server. The conventional ad-hoc network has no printers or servers, only the computers of other users. Two computers have the ability bypass the base station and directly communicate with one another. This also holds true for wireless clients, the main reason why each client must be protected from others. An attacker can easily strike a laptop computer using a wireless connection. For instance, they could initiate a DoS (denial-of-service) attack by jamming wireless clients with illegitimate packets. They can also configure their own client to duplicate the IP (Internet Protocol) or MAC (Media Access Control) address of legitimate clients to disrupt network traffic.

Even if a company only uses hardwired **workstations** and not a wireless connection, a **laptop computer** connected to Ethernet may still have its wireless NIC (Network Interface Card) installed and configurations set in peer-to-peer mode. Wireless NICs send out probe request frames at regular intervals in seek of other devices with the same SSID. This enables a wireless sniffing program to find those devices configured in peer-to-peer mode and attempt to invade the network. From there the attacker could make a connection to the laptop and exploit a number of vulnerabilities in the operating system, possibly gaining administrative access to the machine. With full access they can install a sophisticated Trojan horse or keystroke logging application to further compromise the network. A client-to-client attack may occur when the targeted machine is in transit and in use. At this point, it doesn't matter if the wireless NIC is actively being used or not.

Rogue Access points: RAP (Rogue Access Points) has become a huge issue in wireless security. A Rogue Access Point is one connected to a network without authorization from an administrator. With low-end access points steadily decreasing in price and increasing in availability, RAPs have become much more common. Additionally, many of these access points contain features that make them nearly invisible when coupled with legitimate networks, doing a fine job to conceal their presence.

An intruder deploying a Rogue Access Point in your preexisting wireless network can inflict considerable damage. The use of SSID allows the signal to travel far beyond the exterior of your office building, broadcasting your network traffic throughout the town. An RAP can also be made to appear as if it's one of your subnets. This could easily cause staff members in your network to use them for a connection. When the network client connects with the Rogue Access Point and tries to access a server, sensitive information such as usernames and passwords can be stolen and possibly used in future attacks on the network.

Inception: Any type of wireless communications that is not secured with encryption can be intercepted with a combination of trivial tools such as an inexpensive laptop computer, a wireless network card and free packet sniffing software. The practice of war driving has become increasingly popular as more potential hackers have resorted to driving around in search of insecure wireless access points. Any wireless network using a username and password to allow entrance into a local network is susceptible to interception and traffic monitoring attacks. Many of the sniffing tools used to accomplish this task function by capturing the initial part of the connection, the area that usually includes a username and password. Equipped with these credentials, the intruder can then masquerade as a legitimate user and access the network.

Successful wireless sniffing calls for the intruder to be in close proximity of the targeted wireless traffic. This is typically about 300 feet, although newer wireless equipment is capable of delivering signals much further. At first glance, this feature appears to be advantageous to the user, allowing them to access their network and surf the web further away from the base station. Unfortunately, it also creates a tremendous security risk as intruders can also conduct their attacks from a greater distance. If the intruder is able to sniff out wireless traffic, it is also possible that they can insert false traffic into the connection. From there they can hijack the session by issuing commands on the user's behalf.

Interception Strategies Used to Attack Your Wireless Network: Arpspoof is another popular cracking tool used to exploit wireless networks. It can be easily configured to trick a network into forwarding sensitive information from the backbone directly through the intruder's wireless client. This not only provides them with a way to intercept data, but a way to hijack TCP (Transmission Control Protocol) sessions as well.

A similar method involves tricking a legitimate client into making a connection to a compromised base station set up by the attacker. This is effective because a legitimate user could easily log onto the unauthorized server, sign into fraudulent login screens and unknowingly give away critical data to outsiders. In this scenario, a hacker typically uses a wireless LAN (local area network) program to monitor and intercept information. This type of software gives them the ability to capture both plaintext and encrypted text of a shared key used to authenticate users. After determining the correct response, the intruder can then create a new algorithm using a different exploit to make a network connection as the legitimate user.

Insertion Attack: This practice involves bypassing security mechanisms and inserting unauthorized devices on a wireless network. The device used is usually a laptop or PDA in which the intruder tries to connect it to the base station. Although a base station should be configured to require a username and password to allow access, many times they are not. A more experienced intruder can even physically insert a base station on a targeted network to remotely enable access. Base stations and other wireless devices have become rather affordable and available, giving unscrupulous characters the ammo need to carry out network attacks.

Misconfiguration Attack: The problem arises from the plethora of mismatched software and hardware, making way for a network infrastructure that is vulnerable to a wide range of attacks. In some cases, the devices may function properly but are terribly misconfigured. While several companies take the first step by implementing a security system, many more fail to maintain them, causing these implementations to be inefficient.

Incorporating SSID: SSID (Service Set ID) is a configurable identification mechanism that enables a client to communicate with the correct base station; all stations come included with their own default SSID. When configured properly, only clients configured with the corresponding SSID can interact with the base station. An attacker can exploit the default SSIDs in attempt to access a base station that may have still have its default configuration. Some will change the default SSID password to something simple, ultimately making the network just as vulnerable.

Those configured with more complex SSIDs are still subject to exploits. For instance, an attacker may attempt to guess the base station SSID using a brute force attack of known dictionary phrases, a trick that attempts to guess every word or phrase possible. What may seem like a time consuming process is made easy with the right hacking utilities. The use of simple SSID passwords is considered to be misconfiguration of network resources and makes it much easier for an intruder to compromise a network.

Unlike most of your data, SSID is not encrypted even when enabling the WEP feature, meaning the password is broadcasted in plaintext. This is a major concern as many access points have SSID broadcasting enabled by default. Even if it is turned off, a packet sniffer can simply wait for the next valid user to make a network connection and spy on the plaintext message.

8.5 WIRELESS HONEYPOTS

A wireless honeypot is simply a wireless resource that would wait for attackers or malevolent users to come through on the wireless infrastructure.

Wireless honeypots could help to reveal real statistics about such attacks on your infrastructure, such as the frequency of attacks, the attacker's skill level, his goals and methods. Wireless honepots can also help with protecting your networks while the attacker expends significant effort on fake targets; thus with honeypots blackhats will lose time in their discovery of your network.

8.5.1 Needs and Goals of Wireless Honeypots

Wireless networks are more susceptible to intrusion than their wired counterparts. In wired LAN's access to physical network is easy to protect by firewalls, routers and physical security measures like access badges, etc. But to compromise a wireless LAN's, proximity to other devices is all that is needed. A hacker can be on a nearby floor, or could be outside the building, or could be sitting in a car parked hundreds of feet away from the devices, and still be able to gain access to the network traffic. This lack of necessity of any physical connection makes escape easy and safe for the attacker and, remove any fear from his conscience.

There are still a huge number of open or non secured APs everywhere (hotspots in hotels, airports, public areas, SOHO wireless networks, etc.). The current state of WLAN encryption and authentication is also weak and is a major concern for the Wireless Security.

For corporate hackers and evil cyber-terrorists a breach in wireless network is a easy way to access the wired resources. They can randomly use open APs to anonymously launch attacks, worms, and so on, yet nobody will be able to catch them.

The mail goal is to gather information about real statistics of attacks like network layer being attacked, attack techniques used, the weaknesses of current WEP, WPA and other standards, the frequency of these attacks, the attacker's skill level, his goals and methods.

Wireless honepots can help to detect any intrusion in the network and also consume attacker's time on fake resources. With honeypots blackhats will lose time in their discovery of real network.

8.5.2 Wireless Honeypot History

The concept of Wireless Honeypot was first by Kevin Poulsen in 2002 [10]. He introduced the use of Honeypots in Wireless domain and explained the first organized wireless honeypot.

In June 15th, 2002, Wireless Information Security Experiment, or WISE, was launched by Science Applications International Corporation (SAIC) in Washington DC (US). The focus of this initial research was to get statistics about unauthorized network access, use, and eavesdropping, mostly on open 802.11 based networks.

At the end of 2002, other organizations like Tenebris published the results of collecting data from a wireless honeypot [11] deployed in Ottawa (Canada), and confirmed the huge war driving activity taking place at that time, and the existence of targeted intrusions.

In 2003 [12], and 2004 [13], wireless honeypots were deployed around the city of London and Australia, to investigating the unauthorized use of wireless networks and to promote the idea of using wireless honeypots as a deception mechanism.

In 2004, Laurent Oudot released the article "Wireless Honeypot Countermeasures" [14]. It was focused on providing an introduction to the goals, design and limitations of wireless honeypots, and provided practical examples using honeyd and FakeAP.

In 2006, Wireless Honeypot concept was used in the MAP project [15] (MAP—Measure, Analyze, Protect), in its Measurement component to develop a framework to address existing and future attacks on WiFi networks.

Recently in 2007, Raytheon, sponsored a wireless honeypot research project, dubbed "The Hive" [16], at the University of Florida, to help address wireless threats. The project is based on a Linux.

Case 1: WISE

WISE (2002), Wireless Information Security Experiment, leaded by Rob Lee 1, was the first organized wireless honeypot. Its goal was to gather data about unwary Wi-Fi hackers and bandwidth borrowers, their techniques, attack signatures, frequency, etc. WISE is an "802.11b network" located in Washington D.C. and is dedicated to no other purpose than being hacked from nearby and closely monitors all activities taking place on it. The network has five Cisco access points, a handful of deliberately vulnerable computers as bait, and two omni directional high-gain antennas for added reach to the nearby streets and alleys. On the back-end, a logging host gathers detailed connection data from the access points, while a passive 802.11b sniffer with a customized intrusion detection system acts as a hypersensitive trip wire. It has an Internet connection, hooked up through a Web proxy that intercepts all outgoing connection attempts and presents a consent-to-monitor banner, to get about how Internet link is being used. Like conventional honeypots, the WISE network has no legitimate users, so anything that crosses it is closely scrutinized.

Case 2: KPMG's Wireles Honeypot

KPMG a London based consulting firm set up a wireless honeypot in 2003 to lure London's wardriving commuters. The dummy set-up, designed to appear as a legitimate corporate wireless network, recorded and analysed the activity of users trying to access it. Three separate wireless points were set up at different points around the Square Mile in London, and ran for a week each, aimed at establishing the prevalence of wardrivers and wireless hackers. An average of 3.4 'probes' was detected per working day.

Some interesting characteristics revealed were inferred about the wardriving hackers. Most do it as a hobby, and in some case to use the network to access the Internet. The most popular time for war driving was between 9–10 am, where 24% of probes took place, and 5–6pm where 18% of probes took place. This suggests that people scan for wireless access points while driving in cars, or while on foot or cycling. Virtually no activity was recorded at weekends. So, wardrivers mostly try to access wireless networks on the way to and from work.

Analysis of the probes revealed that 84% of those looking for wireless networks simply identified the presence of the network and moved on. Such probes are expected to be for charting maps of wireless access points for future use. Sixteen percent of probes ended in eventual network access, and three-quarters of those who did access the network undertook activity that would be described as hostile. Deliberately malicious behaviour included attempts to access systems and tamper with their set-up, and attempts to run computer commands that would damage the technology. RSA's survey also found that only a third of the networks detected in their financial district research were running special security technology for wireless networks.

Tim Pickard, Strategic Marketing Director EMEA at RSA Security, comments: "Once again we are seeing how security seems to have been overlooked in the rush to implement wireless solutions. This research clearly demonstrates the very real dangers involved in leaving wireless LANs wide open to potential hackers. CEOs, CIOs and IT Managers need to understand that any investments they have made in securing their infrastructure can be swiftly negated if the backdoor is left open through the introduction of un-secured wireless LANs."

8.5.3 Theory and Design

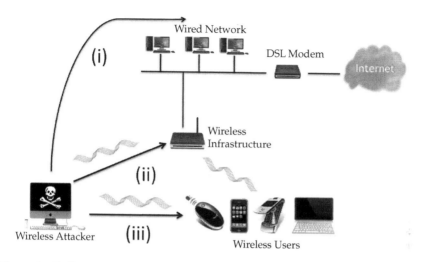

Figure 8.9 Different Scenarios.

There are different attack scenarios to consider when talking about wireless technologies and honeypots, as represented in Figure 8.9:

  i) Attacks directed towards the wired network to which the wireless network connects. These attacks use the wireless network as a medium but the primary target is the network or information systems beyond it.

 ii) Attacks directed towards the wireless users. These attacks may use the wireless network as a medium to target the user's device wireless capabilities, and exploit the fact that the wireless device is enabled. The user may or may not be connected to a wireless network.

iii) Attacks directed towards the wireless network infrastructure. These attacks focus on gaining control of the AP or wireless controllers, that is, the wireless infrastructure devices.

> *Case 3*
>
> **There are very real examples of well known wireless honeypots already deployed: the Science Applications International Corporation (SAIC) created one of the first huge wireless honeypots in Washington DC in order to catch WiFi hackers.**
>
> A project has been conducted by SAIC in Washington DC dubbed WISE (Wireless Information Security Experiment). This implementation used high-end Cisco equipment, high-gain antennas, vulnerable bait systems, logging hosts, a 802.11 sniffer system and a customized intrusion detection system. Unfortunately, SAIC reported that very little interesting activity has been recorded to date.

8.5.4 Wireless Activity

First, attackers try to scan and/or listen to wireless networks, so fake packets can be sent to deceive them, asserting the presence of wireless networks. Or fake wireless resources can be deployed dedicated to some honeypot infrastructure. A very interesting option would be to simulate traffic through the waves of the honeypot, but at this time no automatic or easy-to-use public tool has been released. Automated scripts can be developed to simulate network sessions between an AP and its clients, or use of tools that replay recorded packets such as tcpreplay.

Folks from the French Honeynet Project sometimes use Perl scripts that automate dialogs between clients and servers with random sessions and commands. This idea was first published in June 2003 during the SSTIC in France, by students from ENSEIRB doing some research on UML and Honeypots.

Simulating traffic can be a more important issue on a wireless network dedicated to honeypot activity than on a wired one, because attackers need to see traffic in order to perform some of their attacks. Bypassing 802.1X, bypassing MAC address filtering, cracking malformed WEP keys, looking at beacons, looking at SSID in the frames used for connection by clients, and so on all require existing traffic to be analyzed.

8.5.5 Wireless Architectures

This section discusses Honeypot architectures that can be deployed in wireless environment according to the different attack scenarios [17].

1. Wireless Honeypot Architecture for Attacks against Wireless Beacons

The first honeypot architecture looks for attacks against wireless beacons. Most common types of attacks try to target open networks to exploit their vulnerabilities. Access Point (AP) sends beacons to tell their presence. These beacons contain SSID (Service Set Identifier), time, capabilities, physical layer parameter sets and supported data rates. These features could help an attacker to launch an attack against the network if sniffed.

Figure 8.10 shows Honeypot architecture in such scenarios, where the real network is obfuscated by simulating as many fake networks as possible. Targeting one network is an easy task, but a cloud of targets confuses the attackers and make it difficult for them

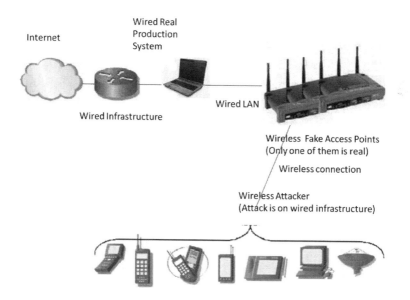

Figure 8.10 Wireless Honeypot Architecture for Attacks against Wireless Beacons.

to attack the real asset. In this way attacks using beacon information are monitored by using FakeAP based wireless honeypots. These honeypots flick from one single SSID to another thus appearing as multiple AP. They transmits fake beacons. This misleads the attacker and traps them into virtual ones.

2. Wireless Honeypot Architecture for Attacks against Wireless Access Point

This Architecture discusses Honeypots for attacks directed towards the wireless network infrastructure. These attacks focus on gaining control of the AP or wireless controllers, that is, the wireless infrastructure devices. These could be open wireless networks with default AP configurations or AP implemented with security specific configurations and with controlling mechanisms. Attackers in such scenario try to gain access of the administrative server of Access Point to get control over the configurations of Access Point. They try to connect to the management interface using well-known default passwords, or try to access other opened services (such as attacks over SNMP, DNS, DHCP, TFTP, etc.) at these server. An attacker could try to use buffer overflow attack to smash the system memory stack and to destroy the whole wireless network.

Honeyd is a tool that can be used to create Honeypot for such scenario (Figure 8.11) by simulating a wireless AP for attackers to connect to. It creates fake TCP/IP stacks to fool remote fingerprinting tools such as Nmap or Xprobe. This could also simulate fake web-servers, fake websites and other fake services to trap the attacker and to hide the real WAP controlling administrative server. Honeypot could also simulate a fake scenario for buffer overflow attack protecting the real memory stack.

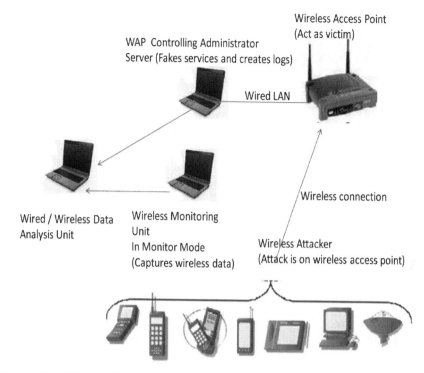

Figure 8.11 Wireless Honeypot Architecture for Attacks against Wireless Access Point.

3. Wireless Honeypot Architecture for Attacks against Wireless Clients

Now a days attackers are very interested in attacking directly the vulnerable wireless client. Such attacks may use the wireless network to which user is connected as a medium to target the user's device. They could also directly attack their wireless capabilities, and exploit the fact that the wireless device is enabled. Attackers listening to the wireless network traffic easily recognize the presence of clients. If clients are not well configured and badly protected (such as laptop used from home and brought to a company), they become interesting and easy targets. An attacker could try to use a Rogue Access Point with a stronger wireless signal than the official wireless AP to lure the wireless clients. They could then easily launch attacks like man in the middle attacks, denial of service, infection with a new worm that spreads itself on the rest of the legitimate network after the client reconnects itself, and so on. Some other attacks against wireless clinets are Wireless driver vulnerabilities exploitation, Wireless client and driver fingerprinting, Wireless 802.11 protocols fuzzing, with and without 802.11 fragmentation, PSPF attacks (direct traffic injection and eavesdropping), Open and WEP-based spoofed access points, Preferred Network List (PNL) attacks, PEAP and TTLS configuration weaknesses.

A honeypot (Figure 8.12)for such scenario would simulate wireless clients that interact with the wireless network. To act as a wireless client, wireless cards are configured in managed mode, with the appropriate settings required to connect to the WAP. It is then deployed at a standard distance of from the WAP. There are multiples options to implement the simulation of multiple clients in an automated way. Client Honeypot creates multiple individual connections against the wireless network, or even remote Internet hosts, simulating the presence of multiple clients using its own unique MAC address. Each client could generate traffic associated to one or multiple protocols (802.1x, ARP, ICMP, TCP, UDP, IPSec, etc.) and applications, such as (secure) web browsing, FTP, SSH, VPN traffic, e-mail access, etc. There are traffic replay tools like tcpreplay, or traffic generators to generate customized traffic for individual client.

Different Levels of complexities can be added to wireless clients as mentioned in HoneySpot paper [18] based on the level of detail and objective tests. A dormat client could be deployed that establishes a connection, exchanges some traffic, and goes to sleep for a few minutes or hours. Such Clients will help to evaluate active and passive session of hijacking through MAC/IP address spoofing in hotspot-like environments. Another option is to pre-design a Preferred Network List (PNL) for the simulated wireless clients to evaluate the occurrence of PNL attacks through access point impersonation. Based on the specific client attacks, different type of simulations can be implemented in wireless clients.

An important to note in wireless client simulations is that Client traffic must be generated in such a way that a casual observer of the wireless network cannot easily determine that the traffic has been automated. The different traffic profiles simulated need to be generated in a random fashion and with varying information data exchanges. If an attacker can easily determine that a client is not a "real" client, for example, by observing that its web browsing pattern is reduced to access to the same site over time, there will be no incentive to compromise it.

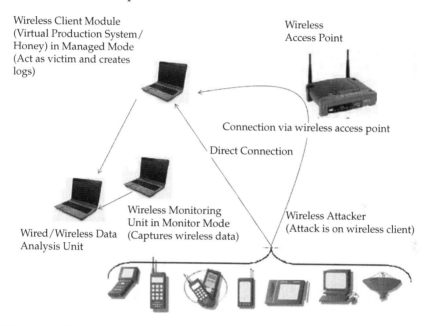

Figure 8.12 Wireless Honeypot Architecture for Attacks against Wireless Clients.

4. Wireless Honeypot Architecture for Attacks against Wired Infrastructure

Figure 8.13 shows Honeypot architecture for attacks directed towards the wired network to which the wireless network connects. These attacks use the wireless network as a medium but the primary target is the network or information systems beyond it.

Honeypot could be made by using a real Access Point to plug it on a wired network (with at least one computer) and with visible resources playing the role of targets on this fake network, and invisible resources to record data and detect intrusions (data capture). Internet access could also be offered to increase the realism of the network. Lance [19] suggest that in such cases the outgoing network traffic needs to be forbidden by using a kind of Intrusion Prevention System, like Snort-inline.

To caught skilled attackers who try to connect to the Internet with tools that encapsulate traffic over DNS, a wireless network with free DNS traffic and requiring authentication for other services (a classic hotspot configuration), could be deployed. Information about the remote IP of the servers attackers are trying to access during their unauthorized tunnel sessions could be getting in this way.

Honeyd can be used to simulate a fake wired network behind the wireless access. Wired devices simulated by honeyd are given virtual IP address and OS signatures are emulated using fingerprinting techniques. It could set up a whole fake Internet routing topology or fake TCP/IP stack.

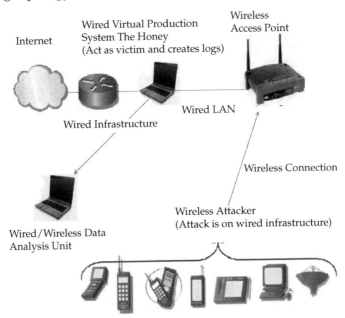

Figure 8.13 Wireless Honeypot Architecture for Attacks against Wired Infrastructure.

5. Integrated Wireless Honeypot Architecture

An integrated Architectures covering all the attack scenarios discussed can be as shown in Figure 8.14. Here we have wireless fake access points to confuse the attacker of real target. Then we have faked web-interface for WAP administrative services. The wired environment behind the real access point is also a simulation to lure the attacker. There are wireless clients also in the network to increase the realism the wireless network and simulates wireless traffic for attacker to sniff. Attack could be anywhere.

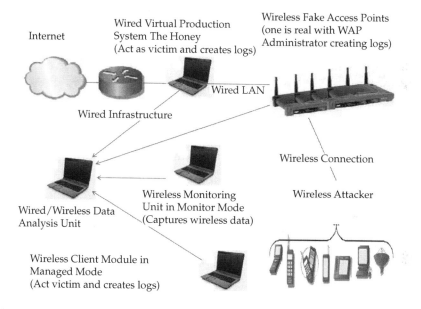

Figure 8.14 Wireless Honeypot Architecture for Attacks against Wired Infrastructure.

8.5.6 Some Practical examples to create Honeypots

First Architecture

For this architecture at least one device is required that offers wireless access. If a real AP is available, then it can be safely plugged onto a wired network (with at least one computer) with visible resources playing the role of targets on this fake network, and invisible resources to record data and detect intrusions (data capture). To monitor wireless-specific layer 2 attacks, data capture can be done on a wireless invisible client in mode Monitor, using software such as Kismet. An example architecture is shown below in Figure 8.15.

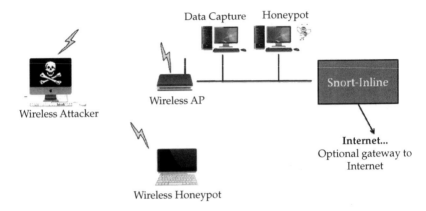

Figure 8.15 Sample WiFI honeypot architecture.

Internet access can be provided on the honeypot network, to improve the realism and interaction of network, however outgoing network traffic should be filtered using a kind of Intrusion Prevention System, like snort-inline from the Honeynet Project to forbid attacks. Most of the time, people don't want to make an Internet connection available to a wireless honeypot because of the related risks however doing so can be used to understand blackhat activities: where do they want to go on the Internet? How do they try do go on the Internet?

For example, if there are only free DNS traffic and require authentication for other services (a classic hotspot configuration), skilled attackers can be caught trying to bounce to the Internet with tools that encapsulate traffic over DNS. Such tools would reveal the remote IP of the server they use to freely access the Internet in their unauthorized tunnel sessions (Nstxd server, for example) which could eventually be used to sue them. If blackhat people were aware of such risks, they would hesitate before doing illegal actions and the impact of wireless crimes would be reduced.

Another option could be the use of wireless clients on such architecture. Usually, people deploying honeypots propose servers, but clients can be used to improve the realism or to monitor specific attacks. More specifically, on a wireless environment, clients can be used to simulate wireless traffic and also monitor layer 2 attacks and probes. In fact, some attackers listening to the wireless network traffic will recognize the presence of clients. Sometimes, those clients are not well configured and badly protected (such as laptop used from home and brought to a company) and become interesting, easy targets. As an example, an attacker could try to use a Rogue AP with a stronger wireless signal than the official wireless AP. A typical client will then automatically connect itself to the attacker's rogue AP and specific, evil actions can then be tried by the attacker: man in the middle attacks, denial of service, infection with a new worm that spreads itself on the rest of the legitimate network after the client reconnects itself, and so on.

Case 4

On 15th December 2009 Project Honey Pot announced that over one billion pieces of spam served to Project Honey Pot and with that milestone they have released their analysis of global spam trends and patterns, and it's *very* interesting.

Project Honey Pot correctly observes that it's actually very difficult to determine the country of residence for a spammer, however it is relatively simple to determine the country of origin of the spam itself. As they explain, "spammers' use of bots can make their messages look like they are coming from somewhere completely different than their actual location. As a result, lists of spam origin countries tell you very little about where the spammers are actually located.... On the other hand, they can help provide insight into a country's security policies because they give evidence on the number of bots operating within a country's borders."

Accordingly, Project Honey Pot has collated a list of the countries with the best IT security—and the worst. In the top 10 list for best IT security, US came in number 6, behind the Netherlands, Australia, Belgium, Canada, and Finland (at number 1). The 10 worst countries for IT security—and thus the ten with the highest number of compromised computers that have been pawned by botnet herders and spammers, and so are being remotely controlled to send spam and other nasty things, include Brazil, Macau, Kazakhstan, Vietnam, Turkey, Macedonia, Columbia and, not surprisingly, China in first place, with South Korea at #3 (Azerbaijan has the dubious distinction of being in second place behind China for having the worst IT security).

Second Architecture

Wireless card can be turned in Master mode to simulate an AP, so that the honeypot is concentrated on only one computer. This is really cheap and easy to manage. Even if the honeypot is compromised, there isn't any problem if it's disconnected from the real network. Moreover, this computer could be either a high-interaction honeypot or a low-interaction honeypot. As an example, a wireless computer (a laptop for testing) with Honeyd can be used, as shown in Figure 8.11

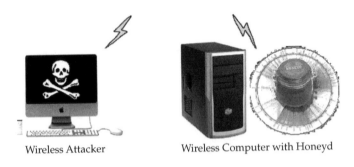

Wireless Attacker   Wireless Computer with Honeyd

Figure 8.16 Simple wireless client in Master mode, with Honeyd.

Third Architecture

A wireless AP can be modified directly and transformed into a honeypot by slightly modifying some classic tools such as Honeyd and compiling MIPS binaries, that would work on this AP and create a very geek, customized, wireless embedded honeypot.

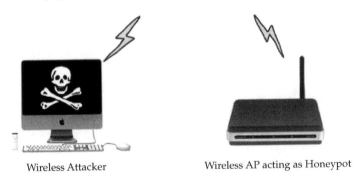

Wireless Attacker   Wireless AP acting as Honeypot

Figure 8.17 Modified AP, hacked firmware and Honeyd.

An additional, and rather evil possibility could be the use of a rogue AP, passively waiting for incoming unauthorized wireless clients, to automatically counterattack them.

8.5.7 Existing Wireless Architectures on Wireless Honeypots

1. WIDS

Wireless Intrusion Detection System [20], [21] incorporates **honeypot** features. It is based on client-server Architecture and has various Modules: honey pot, snort rule analysis, packet analyser, MAC spoofing, web encryption and firewall. Its Honeypot Module creates the environment for honey pot. Whenever any activity takes place on Honeypot it is detected as an attack. Also when an intruder accesses the network of a company, Network using IDS, firewall and other modules identifies the attack. It then diverts attacker to an isolated system created using fake wireless resources like fakeAP and gives unnecessary information to the intruder. **Honeypot** quarantines all the attack events from the production IDS. Honeypot could also detect the jamming of management frames and could decrypt data frames on the fly and re-injected them onto another device. The placement of Honeypot in the overall framework can be seen in Figure 8.18.

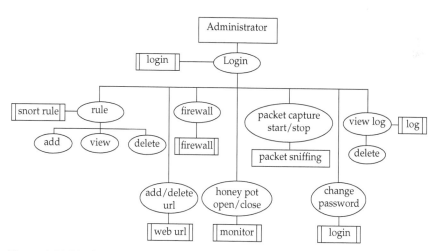

Figure 8.18 Wireless Intrusion Detection System

2. Deceptive Wireless Honeypot

Suen Yek [22] implemented network defence using deception in a wireless honeypot (Figure 8.19). The Deception-in-Depth (DiD) concept is a layered 3 ring architecture with each ring having different deceptive strength. The central core embraces the most effective deception and as the rings progress outward, the strength of the deception abates.

The peripheral, Ring 3, is the most vulnerable FakeAP layer. This layer is similar to the first Generic Architecture. This ring produces an AP gateway for attackers to enter the ring 2. Using FakeAP software, one or many fake access points are simulated to confuse the attacker.

When an attacker is able to find IP address of AP gateway, it reaches the inner virtual wired network simulated using Honeyd. It contains simulated fake web servers and clients workstations accessible to wireless services by doing direct manipulation of the network stack of a designated OS.

The Ring1, the inner most ring is the central logging structure encompassing the IDS SNORT acting as packet sniffer and Honeyd logs to passively record all system traffic. The central is the most important part of the architecture as it has all the network data like the source and destination: IP address, MAC address, TCP/IP ports and the protocols used, as well as any buffer outputs. This collected would be used to confirm network penetration.

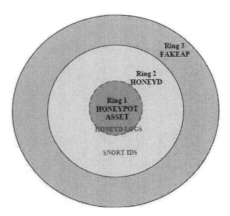

Figure 8.19 Deceptive Wireless Honeypot.

3. HoneySpot

HoneySpot project covers attacks from pure layer-2 attacks in private networks to IP layered attacks in Public hotspot environment, that is, attacks that try to break into a secure wireless network. HoneySpot have defined two types of HoneySpots—Public and Private HoneySpots.

A Public HoneySpot simulates a public wireless data network, that is, pure hotspots networks available at hotels, airports, coffee shops, libraries, as well as other public places where there is a high interest in offering Internet connectivity to visitors and customers. HoneySpot for these networks don't have access control mechanism at the wireless level and focuses on wireless attacks at IP layer, i.e., for "open" networks.

Private HoneySpot simulates a private wireless data network, such as those available in corporations or at home, networks in which access to a wired network (corporate or home network) is offered to only legitimate wireless clients. These networks have access control mechanisms restricting access to the wireless network itself. The honeySpot in this scenario are mostly interested in how an attacker tries to get access to a supposedly "close" network and focuses on pure Layer 2 Attacks.

The architecture provides different levels for both the scenarios. For Public Honeyspot only one level is available Level 0 with Open wireless network (with IP-layer controls). For Private Honeyspot three levels are defined. Level 0 for WEP-based wireless network, Level 1 is a WPA-based wireless network and Level 2 is a WPA2-based wireless network. The Honeyspot architecture have of all the components discussed in Section … i.e., a Wireless Access Point Module, a wireless Client Module, a Wireless Monitor module, wireless data analysis Module, and an optional Wireless Infrastructure module. This is almost similar to the Integrated Architecture discussed in Section .

8.5.8 Wireless Tools

Here are two easy examples for creating wireless honeypots.

Honeyd

- Simulating a network behind the wireless access

  With the help of Honeyd a fake Internet routing topology can be created for a wireless honeypot architecture in order to simulate a huge network on a wireless environment. Such architecture was used during a conference called Libre Software Meeting 2003, where unsuspecting end users connected themselves to a fake network without seeing it was not, in fact, a real one. With such an architecture, an outside attacker could think he has found a big network and would probably lose hours before understanding that it is not.

- Simulating a wireless AP

  One other interesting possibility of Honeyd is the creation of fake TCP/IP stacks to fool remote fingerprinting tools such as nmap or xprobe, and this is an easy way to create your own fake services. For example, by copying well-chosen web pages used to manage an AP, an AP can be simulated. This technique can be useful to monitor attackers who would try to connect to the management interface using well-known default passwords, or who would try other opened services (such as attacks over SNMP, DNS, DHCP, TFTP, etc.).

FakeAP

A novice can use a modem on the phone line to scan remote phone numbers and find open lines like BBSes. This activity was called wardialing, and by transposition in the wireless world, people were talking about wireless scanners or wireless listeners as wardriving, or even warwalking. Wardrivers try to find open networks. A good first idea to delude those potential intruders would be to simulate as many fake networks as possible for them to lose time and patience. Targeting one network is an easy task, whereas dealing with a cloud of targets could be more difficult.

This proof of concept was done with a tool called FakeAP, free software distributed under GPL by the guys from Black Alchemy during the Defcon X. This tool can send specific wireless network traffic to fool basic attackers. As a wardriving countermeasure, it generates 802.11b beacon frames as fast as possible, by playing with fields like BSSID (MAC), ESSID, channel assignments, and so on. This trick is easily created by playing with the tools used to manage a wireless card (under Linux, that's like manually playing with: iwconfig eth1 ESSID RandomSSID channel N...). A remote, passive listener should then see thousands of fake APs! To quote the web site of the authors: "If one AP is good, 53,000 must be better." The idea behind this simple tool was quite good when it was first released, and even NetStumbler users can be detected by looking at 802.11b probe requests/responses. Whereas now, most updated tools can advise the attacker that the detected APs are unusually strange, such as these cases where no traffic is generated on the found networks. Figure 8.20 below, indicates a NetStumbler scan on one of these honeypots.

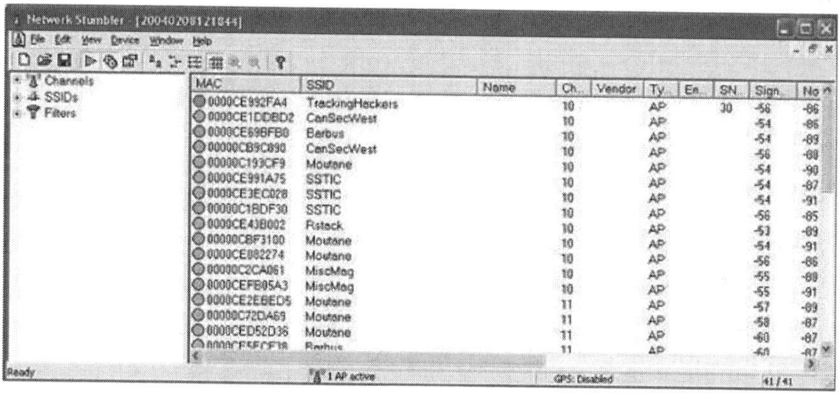

MAC	SSID	Name	Ch.	Vendor	Ty.	En.	SN.	Sign.	No
0000CE992FA4	TrackingHackers		10		AP		30	-56	-86
0000CE1DDBD2	CanSecWest		10		AP			-54	-85
0000CE69BFB0	Barbus		10		AP			-54	-89
00000CB9C890	CanSecWest		10		AP			-56	-88
00000C193CF9	Moutane		10		AP			-54	-90
0000CE991A75	SSTIC		10		AP			-54	-87
0000CE3EC028	SSTIC		10		AP			-54	-91
00000C1BDF30	SSTIC		10		AP			-56	-85
0000CE43B002	Rstack		10		AP			-51	-89
00000CBF3100	Moutane		10		AP			-54	-91
0000CE082274	Moutane		10		AP			-56	-86
00000C2CA061	MiscMag		10		AP			-55	-88
0000CEFB05A3	MiscMag		10		AP			-55	-91
0000CE2EBED5	Moutane		11		AP			-57	-89
00000C72DA69	Moutane		11		AP			-58	-87
0000CED52D36	Moutane		11		AP			-60	-87
0000CE5FCF38	Barbus		11		AP			-60	-87

Figure 8.20 NetStumbler scan on a FakeAP honeypot.

Hotspotter: It passively monitors the network for probe request frames to identify the preferred networks of Windows XP clients, and will compare it to a supplied list of common hotspot network names. If the probed network name matches a common hotspot name, Hotspotter will act as an access point to allow the client to authenticate and associate. Once associated, Hotspotter can be configured to run a command or ascript to kick off a DHCP daemon and other scanning against the new victim. It act as fake access point and runs scripts that simulate simple fake services.

Karma: KARMA is a set of tools for assessing the security of wireless clients at multiple layers. Wireless sniffing tools discover clients and their preferred/trusted networks by passively listening for 802.11 Probe Request frames. Individual clients can then be targeted by creating a Rogue AP for one of their probed networks (which they may join automatically) or using a custom driver that responds to probes and association requests for any SSID. Higher-level fake services can then capture credentials or exploit client-side vulnerabilities on the host.

WiFish Finder: It simulates a virtual wi-fi network envirionment around a probing client. It is a tool for assessing whether WiFi devices active in the air are vulnerable to 'Wi-Fishing' attacks. Assessment is performed through a combination of passive traffic sniffing and active probing techniques. Most WiFi clients keep a memory of networks (SSIDs) they have connected to in the past. Wi-Fish Finder first builds a list of probed networks and then using a set of clever techniques also determines security setting of each probed network. A client is a fishing target if it is actively seeking to connect to an OPEN or a WEP network. Clients willing to connect to WPA or WPA2 networks are also not completely safe from this tools attack.

8.5.9 Wireless Honeypot using Wired Tools

A wireless honeypot system can be made based on the current tools available today.

A standard AP (AP) with logging capabilities, a network sniffer/logger (e.g., snort) and any of the current wired Honeypots like Honeyd can be used. This would give the basic capabilities to monitor wireless hacker activities, and if this is made in a completely stand-alone environment, all of the activity would be considered suspicious activity to be analyzed. Most AP's log, at a minimum, all attempted associations along with the MAC address of the wireless clients. This would provide the traffic patterns and frequency with which attempted access was made. It would be interesting to configure the AP at first with no encryption or security measures in place, and then, incrementally add WEP, MAC address filtering and other security measures to identify the sophistication of the wireless hackers [23]. For this to be successful, it would probably require that this is done in an area with frequent exposure (e.g., near a popular open public WLAN) or by providing some valuable resource that would keep an attacker coming back (e.g., true internet access).

Based on the activity detected on your wired honeypot and network sniffer following can be inferred about the intent of the hacker:

- No activity on the wired honeypot (but association with the AP)—a casual war driver merely cataloging the location (but this could also be for planned future activities).
- Attempts to gain internet access only—a casual wardriver or freeloader looking to gain free internet access (but again this could also be for planned future activities).
- Installation of backdoor or other hacker tools—this might be a more sophisticated hacker looking to compromise a system for future hacking activities.
- Overwriting or deleting key system files—malicious intent to compromise and inflict damage to information systems.

Because a wired network sniffer would only detect an intrusion after the hacker had been associated with the AP, the detection of many suspicious activities (AP scanning, Denial of Service attacks against the AP, spoofed AP's, etc.) would likely be missed thus a wireless intrusion detection and honeypot capability needs to be developed.

With the wireless honeypot using wired tools, all of the necessary honeypot aspects of a wireless honeypot can be simulated (the honey to draw a would be attacker in). A typical hacker is not looking to gain access to the AP, but rather the resources beyond the AP. As of this date, there are no viruses, Trojans, or other malicious code that would reside and operate on an AP. Although, as time passes, the development of new attacks would log all of the associations and traffic passing through a compromised AP or to use the AP to spread malicious code.

8.6 LIMITATIONS

To deploy honeypots to fool attackers, the main aim is to perfectly simulate reality. Many counter papers have recently been released on the Internet because blackhat people want to prove that they are not afraid of honeypots and that they are stronger than their creators. This public game between the good guys and the bad guys will surely improve honeypots technologies, albeit passively, and new paths of research have been drawn to resolve the stealth problems.

Wireless honeypots suffer from the same stealth problems that classic honeypots do, and also from specific, additional ones related to this environment. Remember that skilled attackers may be afraid of "too open" networks. So, the rules of the game are easy:

- The better the simulation of reality, more the number of skilled attackers can be caught (but in this case, intrusions rarely occur);
- The lesser the inclusion of stealthiness, the more successful attacks can be seen (but they are often done by 'kiddies' or inexperienced attackers).

Therefore, depending on goals, one might create honeypots with or without these options:

- Beacon transmission;
- WEP (or more generally, ciphering, that can be cracked more or less easily);
- MAC filtering;
- 802.1X authentication;
- Wireless traffic between clients and AP;
- Wireless clients with auto-connect mode enabled;
- Wireless networks using well known standards (802.11b, 802.11g, 802.11a).

8.7 SUMMARY

It seems that most would concur that the state of wireless security is inadequate, yet, WLAN's continue to be deployed. And, for the home user surfing the web and sending emails or to address certain mobility issues in the corporate world, the current state of security may be adequate if properly deployed. Standards bodies and manufacturers continue to further the security advancement of WLAN's with features like WPA and 802.11i. But, to do jobs as Information Security professional's tools are needed, and a Wireless IDS to support a Wireless honeypot is needed.

This new kind of security resource could easily become an effective way to monitor wireless intrusions attempts and to understand a blackhat's goals and their corresponding tools. Whether these people are corporate attackers, bandwidth borrowers, or cyber terrorists, they will be discovered.

Case 5

SCADA HoneyNet Project: Building Honeypots for Industrial Networks

The short-term goal of this project started by Cisco is to determine the feasibility of building a software-based framework to simulate a variety of industrial networks such as SCADA, DCS, and PLC architectures. The plan is to document the requirements and release proof of concept code (in the form of honeyd scripts) so that a single Linux host can simulate multiple industrial devices and complex network topologies. Given the variety of deployments and the lack of standard, well-defined architectures for industrial networks, this project attempts to create the building blocks so that users can simulate their networks own networks—not make assumptions about what "real world" SCADA/DCS/PLC look like. Assuming deployment of "SCADA HoneyNets" ever reach critical mass, the longer term objective of the project is to gather information about general attack patterns and specific exploits that could be used to write signature for commercial and Open Source IDS products.

EXERCISES

Analytical Questions:

1. What's a "hotspot?"
2. Why is every NIC given a unique MAC address?
3. What is the difference between a wireless router and an access point?
5. How can a wireless card act as an AP?

Short Answer Questions:

1. Explain the Pros and Cons of all 802.11 standards.
2. Explain Infrastructure and Adhoc Modes.
3. Explain the authentication and association procedure in detail.
4. How can IV collisions help an attacker to break into a wireless network?

Long Answer Questions:

1. Give a brief overview of 802.11 technology and standards that relate to equipment connectivity.
2. What are wireless honeypots? Outline the goals of wireless honeypots.
3. Explain the wireless activity on the networks.
4. Give the possible architectures of wireless honeypots.
5. Explain in details the design of wireless honeypots.
6. Write a short note on 1) MITM 2) Jamming attacks.
7. Explain how wireless honeypots can be implemented using wired devices.
8. Enumerate the limitations of wireless honeypots.

Case 6

A little more traffic was recorded by the consulting firm KPMG in London, England. Some interesting characteristics were inferred about the wardriving hackers. Most do it as a hobby, and in some case to use the network to access the Internet… The most popular time for war driving was between 9–10 am, where 24% of probes took place, and 5–6pm where 18% of probes took place. This suggests that people scan for wireless APs while driving in cars, or while on foot or cycling. Virtually no activity was recorded at weekends.

While this may seem harmless, wireless vulnerabilities are significant. Hacker tools are prolific, and the ease at which one can gain access to the traffic should make one ensure that risks are clearly analyzed before implementing a WLAN in any critical business activity. The risk assessment must not overlook the risk of impacting business partner or consumer confidence. In the case of the recent Best Buy incident, they had implemented mobile Point of Sale (POS) Terminals that used 802.11b, but apparently "did not use even the most fundamental security features of WiFi". A researcher was able to sniff the wireless network and capture sales transactions, including what appeared to be credit card information. Best Buy was not completely dependent on their Wireless POS terminals, so business activities did not have to be shutdown, but the incident tarnished their image and may have impacted sales in the immediate time frame. **We must be diligent, we must monitor our wireless networks and we must get smarter about the exploits.**

REFERENCES

[1] "IEEE 802.11," Wikipedia, http://en.wikipedia.org/wiki/IEEE_802.11, 2010.

[2] B. Mitchell, "Wireless Standards-802.11b, 802.11a, 802.11g, and 802.11n The 802.11 family explained," About.com Guide, http://compnetworking.about.com/cs/wireless80211/a/aa80211standard.htm.

[3] "How 802.11 Wireless Works," Microsoft Technet, http://technet.microsoft.com/en-us/library/cc757419%28WS.10%29.aspx, March 28, 2003.

[4] H. Andersson, "Wireless LAN upper layer authentication and key negotiation," RSA Laboratories. January 17.2002: RSA Laboratories, http://www.rsasecuritv.com/rsalabs/technotes.

[5] S. Vibhuti, "IEEE 802.11 WEP (Wired Equivalent Privacy) Concepts and Vulnerability," pp. 1–4.

[6] S. Fluhrer, I. Mantin, and A. Shamir, "Weaknesses in the key scheduling algorithm of RC4," 2001, pp. 1–24.

[7] J.A. LaRosa, "WPA: How it works": compactpci-systems.com, URL:www.compactpci-systems.com/pdfs/Meetinghouse.Apr04.pdf, 2004, pp. 1–5.

[8] L.C. Wong, "An Overview of 802.11 Wireless Network Security Standards & Mechanisms GIAC Security Essentials Certification (GSEC)": sans.org, http://www.sans.org/reading_room/whitepapers/wireless/overview-80211-wireless-network-security-standards-mechanisms_1530

[9] R. Shimonski: windowsecurity.com, www.windowsecurity.com/articles/Wireless_Attacks_Primer.html Jul 30, 2004.

[10] K. Poulsen, "W-Fi Honeypots a new Hacker's Trap", http://www.securityfocus.com/news/552.

[11] E. Jacksch, "Tenebris Wireless Honeypot Project: Assessing the threat against wireless access points. 1.0," CISSP, Tenebris Technologies Inc, 2002.

[12] P. Cracknell, "The wireless honeypot project: A brief look at how wireless networks are used and misused in the City of London," *CISSP Ltd, viewed,* vol. 10, p. 2004, 2003.

[13] P. Pudney and J. Slay, "An investigation of unauthorised use of wireless networks in Adelaide, South Australia," 2005, pp. 29–39.

[14] L. Oudot, "Wireless honeypot countermeasures," *Securityfocus,* http://www.securityfocus.com/infocus/1761, 2004.

[15] "The MAP Project". Dartmouth College. 2006, http://www.cs.dartmouth.edu/~map/

[16] "Wireless honeypots". Randall Brooks. Raytheon. 2007, http://www.raytheon.com/technology_today/current/feature_5.html and http://www.cise.ufl.edu/class/thehive/

[17] R. Goel, A. Sardana, R.C. Joshi "Wireless Honeypots Architectures and Tools", Seminar Report, ECE Deptt. IIT Roorkee, 2010.

[18] R. Siles, "HoneySpot: The Wireless HoneyPot," The Spanish Honeynet Project (December 2007), Spain.

[19] L. Oudot, "Wireless Honeypot Countermeasures," Symantec.com,, http://www.symantec.com/connect/articles/wireless-honeypot-countermeasures 2004.

[20] C. Valli, "Wireless Snort–A WIDS in progress," 2004.

[21] C. Valli and W. Australia, "WITS–Wireless Intrusion Tracking System," 2004.

[22] L.I. Juan and X.U.E. Zhi, "Applying Wireless Honeypot to Deception-in-Depth [J]," *Information Security and Communications Privacy,* vol. 6, 2007.

[23] R. Schoeneck, "Wireless Honeypot," Citeseer, http://www.giac.org/practical/GSEC/Richard_Schoeneck_GSEC.pdf., 2003.

9

Applications of Honeypots

This chapter discusses the practical applications of honeypots for production and research. The chapter begins with the essence of honeypots in defense against automated attack and defense against human intruders. Next their application and advantage over surgical detection methods is discussed. It further delves into role of honeypots in cyber forensics, network surveillance, forensic analysis and tactical battlefield. Finally, its usage as deterrent in production environment has been discussed. The chapter ends with the discussion on application of honeypots in research environment.

Honeypots can be used for productive purposes to provide protection to an organization through prevention, detection and response to an attack [1]. When used for research purposes, they gather information related to the context in which they were implemented. Some organizations study the tendencies displayed by intrusive actions, while others shift their interest towards prediction and anticipated practical applications of honeypots [2] towards network protection and monitoring and prevention.

9.1 DEFENSE AGAINST AUTOMATED ATTACKS

Automated attacks are based on tools which randomly scan entire network, searching for vulnerable systems. If such a system is located, these automated tools attack and take over the system (with worms which replicate inside the victim). One of the methods to protect a system from the automated attacks is to reduce the speed of their scanning activities in order to stop them later on. One of the solutions to these attacks is "Sticky Honeypot". In case of "Sticky Honeypot", it monitors underutilized IP space. When systems are analyzed, Sticky Honeypots interact with those systems and reduce the speed of the attack. This is attained by using a variety of TCP tricks. One way is to set the receiving window size to zero. Reducing its size to zero means that the system has closed its window and cannot receive any more data. So the attacker cannot send regular data packets, refer to Figure 9.1. This technique constantly puts the attacker on hold and is excellent to reduce the speed or prevent the dissemination of worms which have penetrated the internal network.

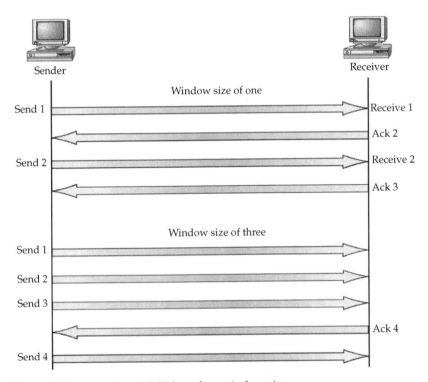

Figure 9.1 Transmission in TCP based on window size.

9.2 PROTECTION AGAINST HUMAN INTRUDERS

Automated attacks may either be accidental or intentional. If these attacks are accidental, they can be prevented in future by applying various patches to cover the system vulnerabilities. However as opposed to automated attacks, human intruders attack the system intentionally with a motive to damage the system or get unauthorized access. Honeypots have been used to protect the systems from human intruders. This concept is known as conning or dissuasion. The idea behind this countermeasure is to confuse the attacker and make him waste time and resources while he is interacting with the honeypot. As the process takes place, the attacker's activity can be detected and the attack can be stopped before it causes any potential damage.

9.3 SURGICAL DETECTION METHODS

Traditionally, detection has been an extremely difficult task to carry out. Technologies like Intruder Detection Systems and Logging Systems have been deficient for many reasons (as discussed in previous chapters): They generate excessive amount of information, inflated percentage of false positives and do not possess the ability of detecting new attacks, work neither in encrypted mode nor in IPv6 environment. Honeypots excel in the field of intrusion detection by solving many of the problems of classic detection. They reduce false positives; capture small amount of data of crucial importance like unknown attacks and new methods to exploit vulnerabilities (zero-days), as well as operate in IPv6 environment.

9.4 CYBER-FORENSICS

Once a network administrator finds out that one of his/her servers was illegally compromised, it is necessary to proceed immediately to conduct a forensic analysis in the compromised system in order to produce an assessment of the damages caused by the attacker. However, there are two problems affecting incident response:

- Frequently, compromised systems cannot be disconnected from the network to be analyzed and;
- The amount of generated information is considerably large, in such a way that it is very difficult to determine what the attacker really did inside the system.

Honeypots help solve both problems due to excellent incident analysis tools, which can be quickly and easily taken off line to conduct a thorough forensic analysis without affecting daily enterprise operations. The only activity traces stored by a honeypot are those related to the attacker, because they are not generated by any other user but the attacker. The importance of honeypots in this setting is the quick delivery of previously- analyzed information in order to respond quickly and efficiently to an incident.

9.5 NETWORK SURVEILLANCE

In their most basic form, honeypots can be thought of as fake information severs strategically- positioned in a test network, which are fed with false information disguised as files of classified nature. In turn, these servers are initially configured in a way that is difficult, but not impossible, to break into them by an attacker; exposing them deliberately and making them highly attractive for a hacker in search of a target.

Finally, the server is loaded with monitoring and tracking tools so every step and trace of activity left by a hacker can be recorded in a log, indicating those traces of activity in a detailed way [3]. The main functions of a honeypot for network surveillance are:

- To divert the attention of the attacker from the real network, in a way that the main information resources are not compromised.
- To capture new viruses or worms for future study.
- To build attackers' profiles in order to identify their preferred attack methods, similar to criminal profiles used by law enforcement agencies in order to identify a criminal's modus operandi.
- To identify new vulnerabilities and risks of various operating systems, environments and programs which are not thoroughly identified at the moment.

As previously discussed, in a more advanced context, a group of honeypots becomes a honeynet, thus providing a tool that spans a wide group of possible threats [4]. This gives system administrator more information for study. Moreover, it makes the attack more fascinating for the attacker due to the fact that honeypots can increase the possibilities, targets and methods of attack.

9.6 FORENSIC ANALYSIS

An attacker may target not only the system resources, but also the system information to damage organization by disrupting its information system. The specific target of the attack may be the system itself or data. Although the attacks which bring down the system are severe and dramatic, they must be well timed to achieve the attacker's goal because immediate and concentrated attention will be applied to restore the system operation. The organizations have to take further security measures so that such attacks cannot be further done.

Storage jamming is an attack very successful in the long run. It installs plausible but incorrect information that misleads an organization into making bad decisions—a technique first identified as storage jamming.

Honeypots can be applied as a useful tool to detect all the attacks on the system and to study a step by step procedure for performing a successful attack of a system [5]. Honeypots help to gather the forensic skills necessary to analyze and learn the threats which the organization faces. Any honeypot server (for example Red Hat 6.0) can be installed and can be used to disguise the attackers. If any sniffer is used, then all actions committed will be captured by the honeypot server. Then the information can be used to find a step by step method used by the attacker to attack the system. Thus honeypots are used widely in forensic researches to get the complete knowledge of attacking style of attacker, their area of interest in the organization, the tools used by them for attacking and many more aspects of attacks.

9.7 TACTICAL BATTLEFIELD

Management of tactical battlefield presents a lot of challenges [6]. The challenges arise due to the volatile and, sometimes, unpredictable nature of these networks. This is because, in a battlefield environment, the network elements (switches/routers) can appear and disappear sporadically, due to jamming, hostile attacks, etc. This results in sporadic changes to the network connectivity. Furthermore, mission re-assignments can occur and may also result in frequent traffic pattern shift. Thus an auto-configuration management system that can adapt to the dynamic network conditions is extremely important. However, the frequent changes and, sometimes, unpredictable (random) nature of the resource limited battlefield networks renders the task of auto-configuration management challenging.

In network deception, attackers are usually intentionally presented with host(s) on the network that has one or more vulnerabilities. Honeypots are network deception tools that are capable of presenting such illusion.

Assuming that the enemy is technically competent and that there are real vulnerabilities, the proper identification of those vulnerabilities by the enemy's intelligence effort will yield a successful attack. Thus, the defensive deception process must be oriented towards defeating the enemy's intelligence process. In a battlefield, deceptions are spread among the normal systems in such a way that the unused services on those systems are consumed with deceptions. This has two effects. One effect is that it spreads the deceptions over a larger portion of the battlefield and the other is that it increases the percentage of deceptions in the environment, thus increasing the likelihood of an intelligence probe encountering a deception rather than vulnerability.

Figure 9.2 can be used to depict the implementation of a battlefield.

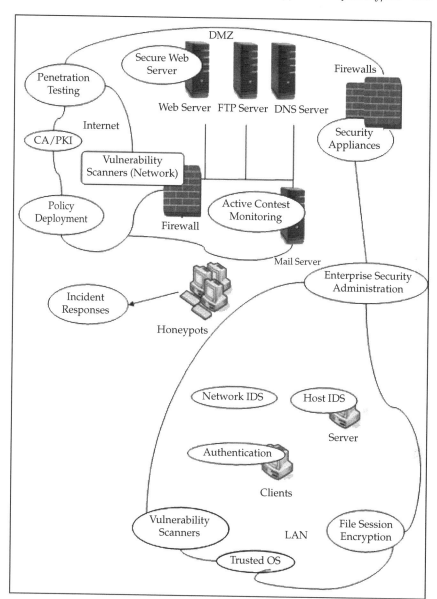

Figure 9.2 Implementation of honeypot in a battlefield.

The depicted Army runs a well set up and maintained security infrastructure with classical elements and recent developments. Services such as web, mail, ftp services and DNS that should be accessible form the outside are situated in a demilitarized zone (DMZ). LAN is in another zone protected by a firewall with adequate, up-to-date security appliances; and even inside the LAN, file transmission is always encrypted. Furthermore, the detection systems: host based intrusion detection systems (IDS), vulnerability scanners and network IDS at the border of the Army's network, are running. The Battlefield might operate virtual private networks (VPN) and a public key infrastructure (PKI) for Intrabusiness Corporation. In this security infrastructure, a honeypot attracts distributed denial of service attackers. DDoS honeypot must fulfill the task to lure the attacker for employing this system as a compromised slave. That is why the attacker's packet should be handled by the honeypot while all other regular packets are forwarded to the legitimate destination (web server, mail server, client, e.g.). So, the honeypot simulates the whole network of the organization to the attacker as shown in Figure 9.3.

Every system in the battlefield might be a honeypot. For example, if the attacker's packets to the web server of the battlefield are detected, the packets go to the honeypot for processing. The reply that the attacker gets, should be indistinguishable from a real reply of the web server.

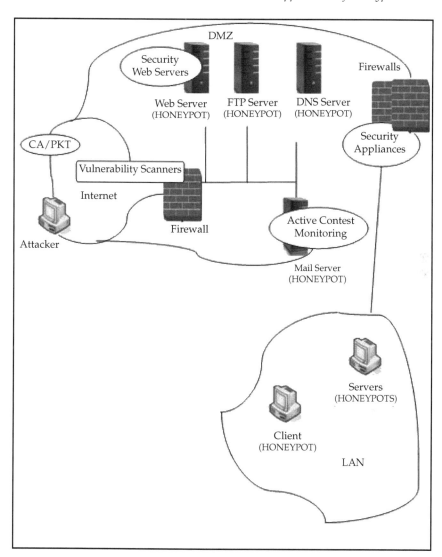

Figure 9.3 View of the attacker.

9.8 USE AS DETERRENT

Honeypots have been used as deterrent. The idea behind using a honeypot as a deterrent is that the vast majority of security breaches are inside jobs performed by trusted employees. That being the case, some companies make it widely known to employees that there are honeypots or even honeynets in place. They can get trapped as the IT department closely guards the identities of real servers and of honeypot servers [7]. The idea behind it is that if an employee is considering hacking into the system, they know ahead of time that the IT department has laid traps for them. Any of the servers on the network could potentially be a honeypot. This might not stop a seasoned hacker, but it would certainly make a lesser skilled hacker think twice before trying to break into an unauthorized portion of the network.

9.9 RESEARCH PURPOSE

Another use for honeypots is that they can be used as a research tool. Hacking techniques are constantly evolving, and the only way that one can hope to keep the network secure is to stay a step ahead of the bad guys. Research honeypots are run by a non-profit research organization or an educational institute. They do not add any direct value to the organization but gather valuable information about all the attacker's activities. This information can be used later to learn the attacking tactics and attacker's motives. Placing a honeypot into a hostile environment allows one to witness the latest hacking techniques first hand so that one can better defend its own network.

9.10 SUMMARY

In the field of security, honeypots have emerged as successful tools and have a wide spectrum of applications. It can be used for defense against automated attacks on one end and for human intruders on the other. The flexibility of honeypots allows them to be used in tactical battlefield. Besides, computer forensics is another area where honeypots are proving to be an excellent tool to analyze past attacks and learn about future attack trends and techniques. Honeypots are being massively used for research purposes.

EXERCISES

Analytical Questions:

1. How do honeypots reduce the speed of the attack?
2. How does the defensive deception process present an illusion to the attacker in tactical battlefield?

Short Answer Questions:

1. What is conning?
2. What advantages honeypots have over Intruder Detection Systems and Logging Systems?
3. Why are honeypots being used in cyber forensics?
4. What are the main functions of a honeypot in network surveillance?

Long Answer Questions:

1. Outline a scenario where honeypots can be used for defense against automated tools.
2. A Tactical Battlefield scenario has been presented in the text where honeypots have been used against attacks. Outline major challenges in tactical battlefield that require the use of honeypots and sketch at least two more scenarios.
3. Which honeypot, production or research, would you recommend for the following cases? (Give appropriate reasons for your choice).

a) Forensic Analysis
b) Tactical Battlefield
c) Network Surveillance
d) Protection from human intruders
e) Protection from automated defense tools

REFERENCES

[1] L. Spitzner, "Honeypots: Definitions and value of honeypots," 2003.

[2] J.W. SONG Fu-qiang , LIU Tao, "Research on Application of Honeypots in IDS," *Modern Computer,* 2008.

[3] T. Yong, L.U. Xi-cheng, H.U. Hua-ping, and Z.H.U. Pei-dong, "Honeypot Technique and its Applications: a Survey [J]," *Journal of Chinese Computer Systems,* vol. 8, 2007.

[4] W. Lu and Q.I.N. Zhi-guang, "Technology and Application of Production Honeynet [J]," *Computer Applications,* vol. 3, 2004.

[5] F. Pouget and M. Dacier, "Honeypot-based forensics," in *Proceedings of Asia Pacific Information technology Security Conference 2004 (AusCERT2004),* Brisbane, Australia, 2004, pp. 1–15.

[6] I. Mokube and M. Adams, "Honeypots: concepts, approaches, and challenges," in *Proceedings of the 45th annual ACM Southeast Regional Conference,* Winston-Salem, North Carolina, 2007, p. 326.

[7] M. Xuan, "Honeypot-Network Trap," *Computer Engineering and Applications,* p. 04, 2003.

10

Anti-Honeypot Technology

This chapter starts by giving overview of current state-of-the-art of honeypot detection by looking at various layers. Some of the issues related to discovering and fingerprinting honeypots are introduced first, and then a few examples such as tarpits and virtual machines. It continues with more practical examples for detecting honeypots, including Sebek-based honeypots, snort_inline, Fake AP, and Bait and Switch honeypots. Further we move to the system world and the application layer and explain how to identify a honeypot by looking at these higher layers. It compares honeypots to steganography and then looks at three common techniques for virtualizing honeypots. For each of these methods, which includes User Mode Linux, VMware environments, and chroot/jail environments, we look at weaknesses that lead to their detection. It continues by discussing the Sebek data capture tool in detail, along with some of the ways it too can be detected. Then we'll discuss some other techniques available for detecting honeypots, such as x86-specific ones and time based analysis. Towards end various techniques for honeypot hunters are discussed. A last the honeypot detection countermeasures have been elaborated.

Honeypots are used to delude attackers and improve security within large computer networks. As this growing activity becomes a new trend, attackers always attempt to defeat the effectiveness of the honeypot. Honeypots are very effective and attackers are working to find ways to exploit and avoid them.

This chapter is to explains how attackers typically behave when they attempt to identify and defeat honeypots. It will help security teams who would like to setup or harden their own lines of deception-based defense. After some theoretical considerations, we will discuss some technical examples to emphasize our explanations.

10.1 NETWORK ISSUES

Attackers might not want to attack a computer being used to trap them, and they might not want to be monitored because this could reveal their identity, their methods and their tools. For example, by using a 0-day exploit against a honeypot that records everything (captures network traffic, low level system activity, and more); an attacker would probably lose the valued secrecy of her techniques. Thus an attacker will try to find is the victim system a honeypot or real system.

Attackers even if get access to a honeypot, for example through a shell or through some custom shellcode, they might still use the network layer in order to determine if they have compromised a honeypot instead of a real machine. By doing this, they might reveal their techniques used to fingerprint a honeypot through the network layer. By then, defenders who operate the honeypot will already have a record of the attacker's malicious activity as a kind of burglar alarm.

Lance Spitzner, founder of the Honeynet Project, said *"a honeypot is an information system resource whose value lies in unauthorized or illicit use of that resource."* So, for example, if a potential attacker is able to determine that a computer is a honeypot without being detected, one might believe it to be a problem. But most of the time, honeypots record almost everything so it is difficult for an attacker to be totally stealthy. Even if a honeypot is not a real computer resource on a production network and it is just sitting there, waiting to be attacked, there are ways to determine its role. Such an activity is called *fingerprinting*. If you want to understand how attackers succeed in fingerprinting attempts against a honeypot, just ask yourself: *what is the difference between honeypot architecture and a real architecture?* Though this might look very easy and simple on the surface, this is the key question to consider when thinking about cloaking and honeypots. Attackers will try to evaluate if this small world attained by them is a real or fake one. This is the problem of determining reality. Depending of the level of interaction and type of honeypot, the fingerprinting methods used may need to be different.

If the operator of a honeypot is trying to emulate an environment, the attacker will probably try to find a path to the truth by looking at specific differences that exist in the compromised envrionment, as compared to a real one. Imagine for example that one deploy a fake proxy service to catch spammers. Consider these questions:

- Will it respond to any possible requests the way a real proxy service would do?
- What if the spammer sends unusual or abnormal requests?
- What if the spammer tries to use unimplemented or complex functions that should be found on your service?
- What if the spammer tries to use the proxy to test if it remains functional under heavy load, and for a long time?

Thus simulating reality is not so easy. In the above example about spammers, aggressors will probably use remote actions based on layer 7, such as sending bogus requests. By using protocols to their practical limits, attackers will probably find a way to fingerprint the simulation of our fake world.

One solution might be the use of a high interaction honeypot that is based on a real system. From a cloaking point of view, a perfect honeypot could be a "sacrificed" one that has a real system installed on it. But when you setup such a security resource, you want to record as many events as possible. Network captures made on the wire of your honeypot might not be enough. Think about attackers who use encrypted channels. It is widely known that some blackhats enjoy SSH sessions to compromised computers. Because of this, people have begun to manipulate the kernel of the operating system in their need to record low-level system events. Such system issues are discussed in next section.

10.1.1 Honeypot and Fingerprinting: Practical Examples

(i) Tar Pits

A tarpit is a computer entity that will intentionally respond slowly to incoming requests. The goal is to delude clients so that unauthorized or illicit use of a fake service might be logged and slowed down. For example, to fight off spammers, some people run tarpits that look like open mail relays, but instead answer very slowly to SMTP commands. These are layer 7 tarpits. Other known tarpits are those that play with the TCP/IP stack in order to hold the incoming client's network socket open while forbidding any traffic over it (layer 4).

The Labrea Tarpit is an excellent example that plays with the TCP/IP stack and has been used to slow down the spread of worms over the Internet, but there are also others such as Honeyd and some native tools in Linux. For example, netfilter/iptables supports a TARPIT target. To achieve this tarpit state, iptables accepts an incoming TCP/IP connection and then immediately switches to a window size of zero. This prohibits the attacker from sending any more data. Any attempt to close the connection is ignored because no data can be sent by the attacker to the target. Therefore the connection remains active. This consumes resources on the attacker's system but not on the Linux server or the firewall running the tarpit. An example iptables rule for TARPIT mode looks like:

iptables -A INPUT -p tcp -m tcp -dport 80 -j TARPIT

Though tarpits are not built to avoid fingerprinting, this is an interesting technical case to propose for our first example.

For a layer 7 tarpit, by looking purely at the latency from the service, an attacker might guess that she has found a fake system after multiple attempts.

For a layer 4 tar pit like Labrea, the TCP window size is reduced to zero, and the tar pit continues to acknowledge incoming packets. This simple signature will probably alert the attacker.

You can see that an attacker (10.0.0.2) trying to reach a fake web server, simulated by Labrea in persistent mode (10.0.0.1), in the following recording made with tcpdump:

03:26:01.435072 10.0.0.2.1330 > 10.0.0.1.80: S [tcp sum ok]

911245487:911245487(0) win 64240 <mss 1460,nop,nop,sackOK> (DF) (ttl 64, id 6969, len 48)

03:26:01.435635 10.0.0.1.80 > 10.0.0.2.1330: S [tcp sum ok]

3255338435:3255338435(0) ack 911245488 win 3 (ttl 255, id 48138, len 40)

03:26:01.435719 10.0.0.2.1330 > 10.0.0.1.80: . [tcp sum ok]

1:1(0) ack 1 win 64320 (DF) (ttl 128, id 4970, len 40)

03:26:01.435887 10.0.0.2.1330 > 10.0.0.1.80: . [tcp sum ok]

1:4(3) ack 1 win 64320 (DF) (ttl 128, id 4971, len 43)

03:26:01.436224 10.0.0.1.80 > 10.0.0.2.1330: . [tcp sum ok]

1:1(0) ack 4 win 0 (ttl 255, id 44321, len 40)

03:26:03.731433 10.0.0.2.1330 > 10.0.0.1.80: . [tcp sum ok]

4:5(1) ack 1 win 64320 (DF) (ttl 128, id 4973, len 41)

03:26:03.731673 10.0.0.1.80 > 10.0.0.2.1330: . [tcp sum ok]

1:1(0) ack 4 win 0 (ttl 255, id 35598, len 40)

By looking at the answers from 10.0.0.1, you will at first notice a window size of 3 and then 0 for the next connection (win 0). You can then understand how an attacker could fingerprint this behavior easily.

Labrea also has the capability of answering requests sent to computers that do not exist. By looking at unanswered ARP requests, Labrea might be configured to simulate unused IP addresses, which is very interesting way to fight worms on large networks with thousands of such IP addresses. If an attacker is on the same network segment as Labrea, there is a way to do fingerprinting at layer 2: this daemon always answers with the same unique MAC address 0:0:f:ff:ff:ff, which acts as a kind of black hole, and thus there is an obvious way to detect it. By looking at such ARP responses, the attacker might have such a concern:

04:59:00.889458 arp reply 10.0.0.1 (0:0:f:ff:ff:ff) is-at 0:0:f:ff:ff:ff

If you want to explore this as an exercise, you can find and change this hard coded value in the sources of Labrea (PacketHandler.c):

u_char bogusMAC[6] = {0,0,15,255,255,255};

(ii) VMWare and Honeypot

VMWare is well known commercial software for virtual machines that allows you to launch multiple instances of different operating systems on a single piece of hardware. It is widely used by honeypot operators because it allows, among other things, an easy deployment of honeypots. Sometimes you can guess that a system is running on top of VMWare by looking at the MAC addresses. It does not mean that this is a honeypot, but this might give pause and some doubts to an aggressor. If you look at the IEEE standards, you will find this current range of MAC adresses assigned to VMWare, Inc:

00-05-69-xx-xx-xx

00-0C-29-xx-xx-xx

00-50-56-xx-xx-xx

So, if one sees such a MAC address either by looking at the cached MAC addresses (via arp -a) or by looking at the data related to the interface (Unix: ifconfig or Windows: ipconfig/all), an aggressor might find something interesting.

Some attackers try to reach remote NetBIOS services in order to launch Windows specific attacks. Honeypots builders dream of catching 0-day exploits against a patched system, but using the Windows integrated firewall might stop most attackers. That's why they often open the related Windows ports (NetBIOS ports, including 135, 137-139 and 445 TCP/UDP), waiting for an intruder. But what if an attacker interacts with the NetBIOS service she will be able to get the MAC address and guess that a system is in fact a VMWare guest (Unix: nmblookup or Windows: nbtstat -A @IP). Some could argue that it is possible to change the MAC address in the configuration of VMWare, but still only some addresses might be accepted: VMWare's MAC addresses are beginning with 00:50:56 (e.g., ethernet0.address = 00:50:56:XX:YY:ZZ).

There are also other points of interests for attackers that would like to fingerprint a VMWare owing to MAC addresses. For example, when the VMWare ESX server automatically generates MAC addresses like 00:05:69:XX:YY:ZZ, it usually means that the IP address of this server is like A.B.C.D where XX is the hexadecimal of C, and YY is the hexadecimal of ZZ. This might reveal the use of NAT before the VMWare box (different external address).

(iii) Sebek based Honeypots

Sebek is client/server based application, and it is the primary data capture tool used by honeynet researchers to capture the activities of an attacker found inside a honeypot. It is a kernel-based rootkit that hijacks the read() system call and it is therefore able to record all data accessed via read(). Sebek lives entirely in kernel-space and has access to all data read, so it is able to access most communication unencrypted. It can, for example, log SSH-sessions, recover files copied with SCP and record all passwords used by intruders. The recorded data is sent in a covert manner via UDP to the Sebek server, the other part of Sebek's client/server architecture. This transmission is done by modifying the kernel in order to hide these outgoing packets such that an intruder cannot see them. In addition, all network counters and data structures have to be adapted in order to make detecting these changes more difficult.

It is possible to fingerprint the presence of Sebek by using the network layer, however, because Sebek records everything which is accessed via read() and then sends this data over the network, some congestion effects will be visible if we read lots of data coming out of the system. When we read a single byte via a read(1) call, Sebek has to transfer nearly 100 bytes of data, including all the network headers, over the network to the host doing the logging. So if we are able to do a read(1) some tens of thousands of times per second, this will lead to a congested network and finally to dropped packets.

We are able to generate many read(1) calls with the help of the dd command:

user@honey:~ dd if=/dev/zero of=/dev/null bs=1

And we are able to identify a congested network with the help of the ping command, as outlined below.

We first ping a local IP-address (for example, the gateway) in order to get an overview of the current congestion of the network. Then dd is started in the background and we run the ping command again. If Sebek is installed on a host, this leads to a significant increase in the average round-trip time. In practical tests, the average round-trip time grew from 0.7 milliseconds to over 4800 milliseconds.

(iv) Snort_inline

Snort_inline is an inline packet modification engine, and it provides for packet rewriting of what looks like something dangerous into something harmless. It is basically a modified version of the popular Intrusion Detection System (IDS) Snort and adds several new rule types (drop, sdrop and reject) to tell iptables whether a packet should be dropped, rejected, modified, or allowed to pass based on a Snort rule set.

An example for this technique is the replacement of the string /bin/sh in shellcode with the string /ben/sh. Another example is the replacement of certain characteristic patterns in network traffic in order to render these attacks useless. This kind of Intrusion Prevention System (IPS) is used in GenII (2nd generation) Honeynets in order to identify and block known attacks. Snort_inline is also necessary because it hampers the possibility of an attacker bouncing from a compromised honeypot to attack further systems. So Snort_ inline also adds some safety from a legal point of view.

But the dropping or modification of outgoing packets leads to an easy way for attackers to spot the presence of Snort_inline: If outgoing packets are dropped, that might look suspicious for an attacker, like a kind of black hole. If outgoing packets are modified, this will be detected by an sophisticated attacker. She will send packets with malicious content to another host which she controls and check whether the packet arrives at the destination in unaltered form (albeit with some small things changed, like a decremented Time To Live). If the content of the packet differs, she will probably conclude that this is a bit fishy. The stealthiness of Snort_inline can be increased with fewer rules, but this leads to more attacks that will not be noticed. A trade-off between stealthiness and safety has to be found.

Moreover, GenII Linux Honeynets typically use the netfilter/ iptables-functionality of the kernel to implement connection limiting. This prevents the abuse of a compromised honeypot for denial-of-service attacks or mass scanning and nevertheless allows an attacker to download toolkits or setup automated bots. As an example, the default configuration of the rc.firewall script provided by the Honeynet Project allows 15 outgoing TCP-connections and 50 outgoing ICMP packets per day.

[...]

Set the connection outbound limits for different protocols.

SCALE="day"

TCPRATE="15"

UDPRATE="20"

ICMPRATE="50"

OTHERRATE="15"

[...]

This connection limiting also leads to a problem with the secrecy of a honeypot: an aggressor can simply try to open many outgoing connections to different hosts or send a lot of ICMP echo request packets to varying hosts. If she notices that after some time no more outgoing packets can reach the target, then this is a clear sign that something weird is going on and that she probably fell for a honeypot. Again, a trade-off between stealthiness and safety has to be found. If you allow more outgoing packets, the detection will be harder, but an attacker has more ways to abuse the honeypot.

(v) Fake AP

Fake Access Point is a Perl-script that generates bogus 802.11b beacon frames with random ESSID and BSSID (MAC-address) assignments. It sends these frames on a random channel and is also able to simulate the wired equivalent privacy (WEP) protocol. A normal access point will "hide in sight amongst Fake AP's cacophony of beacon frames" and therefore this kind of tool can also be used as a wireless honeypot: just deploy one Linux machine running Fake AP near your wireless network and watch for any suspicious traffic. Legitimate users will know the SSID of the network and can connect without problems. Malicious users will try to connect to your network with different SSIDs and can thus be spotted easily.

In its current version, Fake AP does not generate fake traffic on one of the simulated access points and hence there is an easy way to detect the presence of Fake AP: This tool only sends beacon frames and also does not send any real traffic. So an attacker can just monitor the network traffic and easily notice the presence of Fake AP.

(vi) Bait and Switch Honeypots

Traditionally, information security follows the classical security paradigm of "Protect, Detect and React". In other words, try to protect the network as best as possible (such as by using firewalls), detect any failures in the defense (with intrusion detection systems), and then react to those failures (perhaps by alerting the admin via mail). The problem with this approach is that the attacker has the initiative, and she is always one step ahead. The Bait and Switch Honeypot is an attempt to turn honeypots into active participants in system defense. It helps to react faster on threats. To archieve this goal, the Bait and Switch Honeypot redirects all malicious network traffic to a honeypot after a hostile intrusion attempt has been observed. This honeypot is partially mirroring the production system and therefore the attacker is unknowingly attacking a trap instead of real data. Thus the legitimate users can still access all data and work on the real systems, but the attacker is lured away from all interesting systems. As an additional benefit, the actions of the aggressor can be observed and then his tools, tactics and motives can be studied. A Bait and Switch Honeypot is based on Snort, iproute2, netfilter/iptables and some custom code.

An attacker might fingerprinting the presence of a Bait and Switch Honeypot by looking at specific TCP/IP values like the Round-Trip Time (RTT), the Time To Live (TTL), the TCP timestamp, and others. After a switch event, the attacker will stop talking to the real computer, and will start to interact with the honeypots. During the switch from the real system to the honeypot, a sudden change in the IPID can be observed. Previous TCP/IP values will also probably change after the switching has taken place and this can be observed by a sophisticated attacker.

(vii) Commercial Tools

There are even commercial tools such as Honeypot Hunter that use anti-honeypot technology. Honeypot Hunter checks with lists of HTTPS and SOCKS4/SOCKS5 proxies for honeypots, and it is used by spammers in order to detect the presence of tarpits or other kinds of honeypots/proxies. Honeypot hunter works by opening a local (fake) mail server on port 25 (SMTP) and connects back to itself through the proxy. A honeypot is detected if the proxy reports that the connection is up but the tool does not receive a connection to this simulated mail server. This approach identifies most invalid proxies and honeypots and the approach is quite simple. But it can be circumvented easily if you allow a small, but limited, number of outbound connections from the honeypot/proxy. The mere availability of such a program shows that the cyber battle between detection and stealth ness of honeypots has not only begun, but that an arms-race will likely follow.

10.2 SYSTEM ISSUES

As honeypots are being deployed more and more often within computer networks, attackers have started to devise techniques to detect, circumvent, and disable the logging mechanisms used on honeypots. Here explanation of how an attacker typically proceeds as he attacks a honeypot for fun and profit. Several techniques and some diverse tools which help attackers to discover and interact with honeypots will be presented. The section aims to show those security teams and practitioners who would like to setup or harden their own lines of deception-based defense what the limitation of honeypot-based research currently is.

Honeypots versus Steganography

The goal is to hide the existence of a communication channel to anyone but the intended recipient of a message. As an art and science, it came to the forefront a few years ago when Simmons introduced his classic *prisoners problem*. Assume two prisoners are jailed in different cells. A warden has been authorized to carry messages from the one to the other. If the messages are ciphered —which means the warden cannot understand the content of the message—he will become suspicious, and the communication channel will be stopped. But if the prisoners have agreed on a code (for instance, a red sun on a painting is a code to mean something, while a yellow sun means something else), the message will not be noticed by the warden, and the prisoners will have the chance to covertly plot their escape.

When a high interaction honeypot is configured, one hopes to capture a great deal of information about the attacker's activity. Even if he notices he is on a honeypot, learning how he noticed it to be a fake system is still valuable information. So, honeypots do need to be covert, but not necessarily completely covert.

Steganography and honeypots share some characteristics: mainly, that once you are discovered, the game is almost over. Also, in both steganography and honeypots you have to hide the presence of something as best you can. But there are always signs that you leave that inevitably allow for detection. For example, let's use our analogy with the warden again. He may examine the image he's carrying, and if he looks closely he will notice differences between several pictures, and perhaps become suspicious. For honeypots, the situation is comparable: if an attacker carefully watches for signs of deception, he will sooner or later find some.

Since honeypots are being deployed all across the Internet, more and more attackers' tools are starting to include automatic detection of suspect environments. This has already begun with the backdoor-virus-worm known as AgoBot (also known as Gaobot).

Many tools are available for building a high interaction honeypot. We will focus some of the most known, and help show you the inside of the matrix.

Issues in Different Tools

(i) User Mode Linux (UML)

Basically, UML is a way to have a Linux system running inside another Linux system. The initial Linux kernel the *host kernel* (or *host OS*), while the one started by the command linux will be called the *guest OS*. It runs "above" the host kernel, all in userland. Note that UML is only a hacked kernel that is able to run in userland. Thus, you have to provide the filesystem containing your preferred Linux distribution.

By default, UML executes in TT (*Tracing Thread*) mode. One main thread will ptrace() each new process that is started in the guest OS. On the host OS, you can see this tracing with the help of ps:

```
host$ ps a
[...]
1039 pts/6    S    0:00 linux [(tracing thread)]
1044 pts/6    S    0:00 linux [(kernel thread)]
1049 pts/6    S    0:00 linux [(kernel thread)]
1051 pts/6    S    0:00 linux [(kernel thread)]
1053 pts/6    S    0:00 linux [(kernel thread)]
1055 pts/6    S    0:00 linux [(kernel thread)]
1057 pts/6    S    0:00 linux [(kernel thread)]
1059 pts/6    S    0:00 linux [(kernel thread)]
1061 pts/6    S    0:00 linux [(kernel thread)]
1063 pts/6    S    0:00 linux [(kernel thread)]
1064 pts/6    S    0:00 linux [(kernel thread)]
1065 pts/6    S    0:00 linux [(kernel thread)]
1066 pts/6    S    0:00 linux [(kernel thread)]
1068 pts/6    S    0:00 linux [/sbin/init]
1268 pts/6    S    0:00 linux [ile]
1272 pts/6    S    0:00 linux [/bin/sh]
1348 pts/6    S    0:00 linux [dd]
[...]
```

You can identify the main thread (PID 1039) and several threads which are ptrace()d: several kernel threads (PID 1044—1066), init (PID 1068), ile (PID 1268), a shell (PID 1272), and dd (PID 1348).

One of the big issues with UML is that it does not use a real hard disk; it uses a fake IDE device called *ubd**. If you take a look at the /etc./fstab, execute the command mount, or check the directory /dev/ubd/, you will notice that you are inside a UML system. To hide that information, it is possible to start UML with the options fake_ide and fakehd. But don't forget that what you read may not, in fact, be true: have a look at the UML's root device ubd to see that it is 98 (0x62).

In addition, the entries iomen, ioports, interrupts, and many others look suspicious. To counter this way of fingerprinting UML, you can use hppfs and customize the entries in the /proc hierarchy.

Another place to look for UML at is the address space of a process. On the host OS, the address space looks as follows:

```
host$ cat /proc/self/maps
08048000-0804c000 r-xp 00000000 03:01 1058722 /bin/cat
0804c000-0804d000 rw-p 00003000 03:01 1058722 /bin/cat
0804d000-0806e000 rw-p 0804d000 00:00 0
b7ca9000-b7ea9000 r--p 00000000 03:01 171 /usr/lib/locale/locale-
archive
b7ea9000-b7eaa000 rw-p b7ea9000 00:00 0
b7eaa000-b7fd3000 r-xp 00000000 03:01 781848 /lib/tls/i686/
cmov/libc-2.3.2.so
b7fd3000-b7fdb000 rw-p 00129000 03:01 781848 /lib/tls/i686/
cmov/libc-2.3.2.so
b7fdb000-b7fde000 rw-p b7fdb000 00:00 0
b7fe9000-b7fea000 rw-p b7fe9000 00:00 0
b7fea000-b8000000 r-xp 00000000 03:01 782112 /lib/ld-2.3.2.so
b8000000-b8001000 rw-p 00015000 03:01 782112 /lib/ld-2.3.2.so
bfffe000-c0000000 rw-p bfffe000 00:00 0
ffffe000-fffff000---p 00000000 00:00 0
```

In contrast, the address space inside the guest OS looks like this:

```
uml:~# cat /proc/self/maps
08048000-0804c000 r-xp 00000000 62:00 9957 /bin/cat
0804c000-0804d000 rw-p 00003000 62:00 9957 /bin/cat
0804d000-0806e000 rw-p 0804d000 00:00 0
40000000-40016000 r-xp 00000000 62:00 13907 /lib/ld-2.3.2.so
40016000-40017000 rw-p 00015000 62:00 13907 /lib/ld-2.3.2.so
40017000-40018000 rw-p 40017000 00:00 0
```

4001b000-4014b000 r-xp 00000000 62:00 21846 /lib/tls/libc-2.3.2.so

4014b000-40154000 rw-p 0012f000 62:00 21846 /lib/tls/libc-2.3.2.so

40154000-40156000 rw-p 40154000 00:00 0

9ffff000-a0000000 rw-p 9ffff000 00:00 0

beffe000-befff000—p 00000000 00:00 0

What one should notice, and what is not that common, is the topmost address which indicates the end of the stack (forget about the mapping of the dynamic libraries). Depending on the amount of memory available on your host, it is usually 0xc0000000. However, on the UML, we have 0xbefff000. In fact, the address space between 0xbefff000 and 0xc0000000 on a UML contains the mapping of the UML kernel. This means that each process can access, change, or do whatever it wants with the UML kernel.

(ii) VMware

VMware is a very efficient virtual machine which provides a virtual x86 system. Thus, you can install (almost) any Operating System you want, from Linux or Windows to Solaris 10.

The first step to detect a VMware is to look at the hardware that it is supposed to emulate. Prior to version 4.5, there were some specific pieces of hardware that are not configurable:

- the video card: VMware Inc [VMware SVGA II] PCI Display Adapter,
- the network card: Advanced Micro Devices [AMD] 79c970 [PCnet 32 LANCE] (rev 10),
- the name of IDE and SCSI devices: VMware Virtual IDE Hard Drive, NECVMWar VMware IDE CDR10, VMware SCSI Controller.

It is possible to patch the VMware binary to change these default values, however. Kostya Kortchinsky from the French Honeynet Project has written such a patch, which is able to set these values to some other values. This patch is publicly available.

Furthermore, the VMware binary also has an I/O backdoor. This backdoor is used to configure VMware during runtime. The following sequence is used to call the backdoor functions:

MOV EAX, 564D5868h; Magic Number

MOV EBX, COMMAND_SPECIFIC_PARAMETER

MOV ECX, BACKDOOR_COMMAND_NUMBER

MOV DX, 5658h; Port Number

IN EAX, DX

At first, register EAX is loaded with a magic number that is used to "authenticate" the backdoor commands. Register EBX stores parameters for the commands. In register ECX the command itself is loaded. The following table gives an overview of some possible commands:

Command Number	Description
00h..03h	?
04h	Get current mouse cursor position.
05h	Set current mouse cursor position.
06h	Get data length in host's clipboard.
07h	Read data from host's clipboard.
08h	Set data length to send to host's clipboard.
09h	Send data to host's clipboard.
0Ah	Get VMware version.
0Bh	Get device information.

In total, there are at least 15 implemented commands.

Register DX stores the I/O backdoor port, and with the help of the IN instruction, the backdoor command finally gets executed. It is clear that with the help of the VMware I/O backdoor it is possible to interfere with a running VMware.

With the help of Kostya Kortchinsky's patch, you can change the magic number and thus "hide" the backdoor from an attacker.

(iii) Detecting additional lines of defense: chroot and jails

chroot() was never designed for security, but it is considered to be a necessity when one wants to protect a sensitive server. Detecting that you are in a chroot environment , or even circumventing it, is not really that difficult.

Unless the chroot directory is on a specific partition, and placed at the top of it, the inode numbers are not those expected of a real root directory:

ls -ial /

2 drwxr-xr-x 24 root root 4096 2004-11-30 08:14 .

2 drwxr-xr-x 24 root root 4096 2004-11-30 08:14 ..

...

Here, the directories inodes of . and .. are the same, and are equal to 2 (which is the normal value for a root directory on a partition). In the current directory, we have:

ls -ail.

1553552 drwxr-xr-x   6 raynal users   4096 2004-12-14 13:58 .

6657574 drwxr-xr-x   6 raynal raynal 4096 2004-12-12 16:25 ..

Then, when we chroot a shell in the current directory, we retrieve the same inodes numbers:

chroot . /bin/busybox

BusyBox v0.60.5 (2004.10.29-22:08+0000) multi-call binary

ls -ial

1553552 drwxr-xr-x   6 1000   100   4096 Dec 14 12:58 .

1553552 drwxr-xr-x   6 1000   100   4096 Dec 14 12:58 ..

While the .. has been changed to match the . directory, it is still not the expected value.

Note that there is much more to do in a chroot. For instance, you can send signals to any process outside the chroot(), or even attach to outside processes with ptrace(). Since ptrace() can be executed from inside the chroot on any process that is outside the chroot(), the attacker has an easy way to inject whatever he wants on the host. Such evasions are also possible through mount(), fchdir(), sysctl() and so many others.

When we think about virtual environments and security, it's pretty clear that chroot() is definitely not something to rely upon. Another option to enforce confinement provided by FreeBSD, which is based on chroot() but is more reliable, is the jail(). A jail() let you create a virtual host, bound to an IP address, with its own tools, users, and more. It is very convenient for virtual hosting, and it could be used for honeypots too.

However, even though FreeBSD's jail() is more reliable, it is not really much more covert. There are several tests one can perform to detect if you are in a jail:

- All processes in a jail have a specific 'J' flag, as shown below:

  jail# ps

  PID TT STAT TIME COMMAND

  6908 p0 SJ 0:00.02 /bin/sh

  6910 p0 R+J 0:00.00 ps
- You can also have a look at the PIDs as they do not increase in the usual way.
- The inode number of the root directory is not 2 as expected.

- By default, raw sockets are forbidden as you can see:

  jail# ping -c 3 miscmag.com

  ping: socket: Operation not permitted
- Note that it is now configurable in the latest version of FreeBSD.
- Sniffing in a jail gives access to all the traffic that comes through the device. This is normal since a jail is usually built as an alias on a real device.
- There are surely more ...

(iv) Detecting Sebek

In a previous section we showed that it is possible to detect the presence of Sebek using the network layer. Now we want to concentrate on system level issues with Sebek detection. We will first take a closer look at the Linux version 2.1.7 and then later show similar results for Sebek on other architectures.

Detecting Sebek on Linux

The Linux version of Sebek is a loadable kernel module and works similar to a rootkit. It overwrites the pointer to the read() system call in the system call table with a pointer to its own version. It is therefore able to access all data passed to the read() function and can then monitor this data.

Sebek hides its presence the same way that adore, a popular rootkit, does: it uses cleaner.c, a simple loadable kernel module that unlinks the last module in the kernel's module list. This unlinking prevents programs like lsmod to display the presence of Sebek on a host. But the "hidden" Sebek module still has a complete module header structure in memory. This structure can be found with very little effort and can be used to locate Sebek's code (such as its cleanup_ module() function) and its variable storage. The tool module_hunter.c is able to locate hidden modules. It searches for patterns of module structures in the kernel's address space and is able to detect modules hidden via unlinking, as shown below:

lsmod

Module	Size Used by	Not tainted
ds	7092	2
yenta_socket	10912	2
pcmcia_core	42912	0 [ds yenta_socket]
ipsec	268100	0
usbcore	64972	1

insmod module_hunter.o

cat /proc/showmodules

dmesg | tail -n 9

address	module
0xe0e6d000	usbcore size: 0xfdcc
0xe0e7e000	snd-pcmcia size: 0x50b4
0xe0e86000	ipsec size: 0x41744
0xe0ec9000	module_hunter size: 0x3ec
0xe0f2e000	pcmcia_core size: 0xa7a0
0xe0f3a000	yenta_socket size: 0x2aa0
0xe0f44000	ds size: 0x1bb4

If you compare the above output of lsmod with the output of module_hunter, you will notice two differences: The module with the name "module_hunter" is new because we have loaded this module. And the module "snd-pcmcia" is hidden and thus not displayed by lsmod, however module_hunter is able to find it.

Furthermore, it is possible to locate the address of Sebek's cleanup_module() function using this technique. With the knowledge of this address, disabling this piece of software on a honeypot is quite feasible. A simple kernel module that jumps to the memory location of cleanup_module(), and thus executes this function, is able to remove Sebek from the host. This works because Sebek reconstructs the pointer to the original read() system call (ord in the following code snippet) as shown below:

if(sct && ord){

sct[__NR_read] = (unsigned long *)ord;

}

Now, after calling cleanup_module(), the system call table has its original layout and no further logging takes place.

A technique that is commonly used by rootkit detection tools can also be used to detect the presence of Sebek on a host. By looking at the system call table and analyzing the pointers to the various system calls, it is possible to detect a modified host. In an unmodified system call table, the pointers to the read() and write() system calls are adjacent. Because Sebek changes the pointer of the read() system call, this adjacency is no longer given. Thus analyzing the pointers of this two system calls can detect a modified system call table. An example of this is shown below:

Unmodified system call table:

sys_read = 0xc0132ecc

sys_write = 0xc0132fc8

After loading Sebek:

sys_read = 0xc884e748

sys_write = 0xc0132fc8

You can see that the pointer to the read() system calls points to a far distant memory location and can thus conclude that someone modified the system call table.

Sebek also modifies the network stack to hide its presence. It has to adjust some counters to conceal the transmission of the logging data. This modification can also be detected by a sophisticated attacker who compares the kernel's internal network counters and the output of ifconfig or other tools.

Detecting Sebek on other Architectures

Similar methods of detection are also applicable for Sebek on other architectures. First we will take a look at the Windows version. Sebek for Windows is a kernel driver that hooks all console and GUI read() operations to capture all data. By traversing PsLoadedModuleList, it is possible to detect if a kernel driver is "hidden". This is comparable to searching through memory on Linux looking for "hidden" modules. With the help of tools like KProcCheck it is possible to detect Sebek on Windows, as seen below:

C:\>kproccheck -d

KProcCheck Version 0.1 Proof-of-Concept by SIG^2 (www.security. org.sg)

80400000 - \WINNT\System32\ntoskrnl.exe

80062000 - \WINNT\System32\hal.dll

F7410000 - \WINNT\System32\BOOTVID.DLL

[...]

F7298000 - SEBEK.sys

[...]

Another way to detect Sebek is to look out for hooked APIs, a technique that is similar to looking for memory locations that have changed in the system call table on Linux. KProcCheck is also able to display the Native APIs that are hooked, as we can see:

C:\>kproccheck -t

KProcCheck Version 0.1 Proof-of-Concept by SIG^2 (www.security. org.sg)

Checks SDT for Hooked Native APIs

ZwClose	18 SEBEK.sys [F729A092]
ZwCreateFile	20 SEBEK.sys [F729A98C]
ZwCreateKey	23 SEBEK.sys [F729AD10]
ZwEnumerateKey	3C SEBEK.sys [F729AE02]
ZwEnumerateValueKey	3D SEBEK.sys [F729AA50]
ZwOpenFile	64 SEBEK.sys [F729A8E6]
ZwOpenKey	67 SEBEK.sys [F729AD88]
ZwQueryDirectoryFile	7D SEBEK.sys [F729A4CC]
ZwQuerySystemInformation	97 SEBEK.sys [F729A5F0]
ZwReadFile	A1 SEBEK.sys [F7299CF0]
ZwRequestWaitReplyPort	B0 SEBEK.sys [F7299F14]
ZwSecureConnectPort	B8 SEBEK.sys [F7299FE6]
ZwWriteFile	ED SEBEK.sys [F7299D48]

Number of Service Table entries hooked = 13

Sebek on OpenBSD

As a last example, let's take a quick look at the OpenBSD version of Sebek. On OpenBSD, Sebek is a patch for the kernel and the functionality is similar to other Sebek versions. All operations that are passed through read() is recorded and sent in a covert manner to the logging host. But once again, the presence of Sebek can be detected by an attacker. By simply searching through the kernel file and disassembling "dofileread" it is possible to detect the presence of Sebek:

echo "disassemble dofileread" | gdb -q /bsd | grep sebek

0xd01c9bdc <dofileread+292>: call 0xd01c967c <sebeklog>

Furthermore, detecting Sebek via bpf fingerprinting and reconstructing sensible information about Sebek is also possible. You can read more about the system issues with Sebek on OpenBSD in a short study by the Droids Corporation.

(v) Other places to look for honeypots

Many people looking for honeypots have been interested in detecting the presence of one inside a virtual machine. While some detection tricks are specific to certain implementations, others depend on the nature of what we want to do with the high interaction honeypot.

Processor

One way to detect the presence of a virtual machine is specific to an x86-based processor. In protected mode, all memory accesses pass through either the "global descriptor table" (GDT) or "local descriptor table" (LDT). The GDT and LDT contain segment descriptors that provide the base address, access rights, type, length, and usage information for each segment. The GDT is used by all programs whereas the LDT can optionally be defined on a per-task basis to expand the addressable range of the specific task.

The "interrupt descriptor table" (IDT) is similar to the GDT and LDT, but it holds gate descriptors that provide access to interrupt and exception handlers. It is thus a direct replacement for the interrupt vector table used in x86-based systems.

The registers GDTR, LDTR and IDTR all contain the linear addresses and sizes of their respective tables. With the help of the three instructions SGDT, SIDT, and SLDT, the contents of these registers can be retrieved.

The SGDT instruction stores the contents of the GDTR in a 6-byte memory location. The SLDT instruction stores the segment selector from the LDTR in a 16 or 32-bit general-purpose register or memory location. The SIDT instruction stores the contents of the IDTR in a 6-byte memory location.

These instructions are normally only used by operating systems but are not privileged in the Intel architecture. This means that these instructions can be executed by a non-priviledged process running in user-land (ring 3). Since an Intel processor only has one LDTR, IDTR, and GDTR, a problem arises when multiple operating systems try to use the same registers—if two operating systems are running concurrently (one host OS and one OS inside a virtual machine), then the host OS must relocate the IDT of the guest OS so that both do not conflict. But a process running inside a virtual machine can execute the SIDT instruction to retrieve the content of IDTR. Since this instruction does not require privileges and does not generate an exception, the host OS can not prevent this instruction. But the process running inside the virtual machine gets the relocated address of the IDT and can conclude that it is running inside a virtual machine.

Time

The best solutions for building a high interaction honeypot are to do with improving logging or using a virtual machine. These techniques do have a price, however. Each action performed by the intruder on the honeypot is *longer* than on a sane system, whereby *longer* can have several meanings:

- More instructions are executed, either to log the true instruction, or to emulate it.
- More time is needed to execute the true instruction, because it is not the only instruction to be executed.

Hence, having reliable ways to measure either the number of instructions or the execution time also provides an efficient way to detect a hazardous honeypot environment.

Delalleau gave a solution to make the time comparison more accurate, by doing many measurements so that they are statistically valid. During a learning phase, one evaluates the distribution for execution times. Then, when one arrives on a suspect host, he can compare the reference distributions between the sane system and the suspect one. To compare distributions, Delalleau used the usual scalar product between the Fourier transform from both sets of measures (one needs to remove high frequencies to suppress the noise). The normalized scalar product of the sets will be close to 1, and will thus indicate a clean host (that is, if the reference set corresponds to a clean system of course), while it will be close to 0 otherwise.

This works quite well for detection, as root-kits hook functions and virtual machines each add several steps in the normal execution flow. Let's see two examples of this:

1. Detecting Sebek by measuring the execution time of the read() syscall is possible. On a system without Sebek, minimal time is around 8225 and scalar product 0.776282, whereas with

Sebek, minimum time is 29999 and scalar product 0.009930. An explanation for this is that read() with Sebek is probably that long because each read() operation causes the sending of at least one UDP packet.

2. Detecting VMware is also very easy, and doing so causes an illegal instruction. This is because VMware's exceptions handler needs to check whether the exception must be handled by VMware itself or by a specific handler present in the process which caused the illegal instruction. The timing to handle an illegal instruction is 776mms on a sane system, while it increases to 2530mms on VMware.

10.3 TECHNIQUES FOR HONEYPOT DETECTION

Spammers continually scan the Internet for open proxy relays by using these open relays, they can obscure their originating IP address and remain anonymous. However, when a spammer comes across a service on a honeypot, that honeypot can collect valuable information about the spammer's true identity and help unmask it.

In response to the threat that honeypots pose to spammers, the first commercial anti-honeypot technology "Send-Safe's Honeypot Hunter" attempts to detect "safe" proxies for use with bulk-mailing tools. This honeypot-detection system's appearance, in association with other emerging spam tools, suggests three important trends:

- honeypots are affecting spammers,
- current honeypot technology is detectable, and
- more honeypot-identification systems are likely.

  Also the ability to detect a honeypot will not be limited to spammers and other hostile or malicious groups could benefit from similar identification systems.

Basic Honeypot Services

Honeypots are designed to resemble valid systems. They use this cloak to collect information about attackers and their methods. To appear as a tempting target, honeypots offer a variety of seemingly vulnerable services. Although the complexity of honeypot services varies dramatically, they generally fall into one of four types: minimal, restricted, simulated, and full. From low complexity to high.

- Minimal servers provide an open service port.
- Restricted servers provide basic interactions.
- Simulated servers provide complex interactions.
- Full servers provide full functional support.

Some minimal servers will reply with a basic connection header, but they usually don't perform anything more detailed. An example of a minimal service is the SMTP server that the BackOfficer Friendly (BOF) honeypot provides: it simply disconnects with the message, "503 Service Unavailable." By adding a minor amount of interaction to a minimal server, a restricted service can appear fully functional, even though no authorization is available. The BOF telnet server, for example, prompts for a username and password, but no valid log in mechanism exists. A simulated service appears as a full working server, but in reality, it logs actions instead of performing external operations. Simulated servers accept log ins and requests, and generate well-known replies and error messages. Examples of simulated servers include scripts that emulate full SMTP and Microsoft IIS Web servers. In contrast to these pseudoservices, full honeypot services are rare. They not only manage requests, but they also let malicious entities fully interact and even compromise the simulated system. Many full honeypots also permit limited external connections, which makes the service appear fully functional while preventing it from taking part in denial-of-service (DoS) attacks.

10.3.1 Honeypot Hunters

Spammers View of Honeypot

Spam developers are generally reactive, not proactive: they only change their tools when those tools become ineffective. For example, one of the first technologies to prevent spam used hash-based filters that summarized each email message's content into a hash table. Repeated hash table entries denoted identical message content—that is, a bulk mailing. To counter hash systems, spam developers created "hash busters"—unique strings that generate different hash values. Similarly, today's bulk mailing tools use anti-Bayesian encoding methods—such as random words, sentences, or paragraphs—to pass Bayesian filters.

Send-Safe Honeypot Hunter is a tool designed for checking lists of HTTPS and SOCKS. Honeypots are fake proxies run by the people who are attempting to frame bulkers by using those fake proxies for logging traffic through them and then send complaints to ones' ISPs. The Send-Safe tool suite makes an extensive collection of bulk advertising tools commercially available. Its bulk mailer is popular for generating spam email, and its proxy scanner can find multiple open proxy servers for obscuring a spammer's identity. Its other tools include an email verifier and a tool for generating bulk instant messages. Send-Safe's latest tool, Honeypot Hunter, suggests that spammers are aware they need to identify honeypots.

One can safely assume that Send-Safe's users are not the only people negatively affected by honeypots. The appearance of this honeypot detection application implies a reactive technological escalation.

Honeypot Hunter's detection methods are likely widely known in the underground community. The free honeyd project has a default installation with fixed-response messages; administrators who don't change the default messages might unknowingly provide an attacker with a unique method for identifying the honeypot. Other detection methods—such as known application error handling, operating system fingerprinting, TCP sequence analysis, and ARP addresses—could also identify a honeypot.

Working of honeypot Hunters

Honeypot Hunter is designed to test open proxy connectivity. Depending on the type of connection response, it classifies the proxy as safe (good), bad (failed), or a trap (honey honeypot).

For example, Honeypot Hunter can test port 1080 for Socks4 and Socks5 proxy support and all other ports for HTTP "CONNECT" proxy support. Honeypot Hunter essentially performs a series of simple tests. First, it opens a false mail server on the local system (port 25) to test the proxy connection and then connects to the server's proxy port. After connecting, Honeypot Hunter attempts to proxy back to its own false mail server. The basic approach of connecting back to itself is enough to identify most invalid proxies and honeypots. In particular, if the remote server claims to have successfully connected, but Honeypot Hunter's false mail server didn't receive a connection, then the proxy is likely a honeypot.

The Effect Honeypot Hunters on Honeypots

The appearance of mainstream honeypot-detection systems has significant ramifications for honeypots. If malicious users can detect honeypots, then they can bypass detection. At minimum, this ability lowers the value of the information gathered because bypassed honeypots would not detect any new attacks. More importantly, if people can detect a honeypot, they can attack it. Three basic approaches exist for attacking a honeypot: compromising, poisoning, and studying. Alternatively, the entity could use the honeypot to stage attacks on other systems throughout the Internet. Instead of compromising the honeypot, a malicious user also could opt to flood the honeypot with false information. This poisoning effectively buries any valuable information under a mound of noise. By poisoning the honeypot, other hostile activities could go unnoticed. Bypassing prevents the honeypot from collecting information and flooding obscures collected information, but an attacker could choose to use the honeypot as it's intended: for gathering information. Just as a honeypot provides valuable insights about an attacker to the observer, an attacker that compromises a honeypot could learn a lot about the observer. He or she could identify personal information such as people's names, operating hours, or skill levels. A compromised host could identify the protected network's organization, items that the organization considers "valuable," and where this value is kept. A compromised Honeyd system that emulates only Windows systems, for example, would suggest a company that only uses Windows; a honeypot database server that emulates Oracle would suggest a corporate Oracle database.

Detection of the Honeypot Hunter

An evaluation of Honeypot Hunter version 1.0 (released on 11 November 2003) and version 1.0.1 (released on 4 December 2003) was performed. Unfortunately, version 1.0 appears unstable and crashes when the false mail server is activated. Moreover, it doesn't seem to generate valid Socks5 requests. Version 1.0.1 corrected the crashing problem but still has issues around email header formatting and honeypot detection.

But, the updated release indicates that Honeypot Hunter is actively and rapidly being developed. Honeypot Hunter provides a significant amount of insight into the honeypot-detection approach. By detecting it, a honeypot could impersonate a full proxy and remain undetected.

However, Honeypot Hunter has many identifiable aspects, including network connection methods, server identification, and test email formats. Honeypot Hunter generates self via-proxy connections, from the Honeypot Hunter system to the proxy and back to itself. A honeypot configured to permit self-via-proxy connections could appear viable to the tool while remaining undetected. A Honeypot Hunter-blocking honeypot could initiate the full proxy connection and determine the mail server type. A server with a different identification would denote a non- Honeypot Hunter system and not need to deliver email sent through the honeypot.

Honeypot Hunter's test email has a fixed number of headers in a specific order with specific capitalization. However, different email programs generate different email headers. A honeypot mail server that only passes messages with these specific headers would be undetectable by Honeypot Hunter:

From: %s

Message-Id: %s%s

Date: %s

Subject: %s

To: %s

Content-Type:

text/plain;

charset="iso-8859-1"

Content-Transfer-

Encoding: 7bit

Although these anti-detection approaches might work with Honeypot Hunter's current version, they are unlikely to be general enough for future honeypot-detection systems. Future honeypot-detection systems will likely use additional detection techniques, different test email formats, and even a variety of test server configurations.

Honeypot systems that use anti detection techniques will likely lead to anti-anti-detection systems. The next logical step for Honeypot Hunter, for example, would be to split the false mail server from the Honeypot Hunter client. This change would remove the self-via proxy connections, thus permitting connections from the Honeypot Hunter client to the proxy to a different Honeypot Hunter server under the user's control. Removing the self-via-proxy connections would make detecting the anti-honeypot system more difficult. Moreover, Honeypot Hunter's client component could relay tests through known open proxies and hide the user's true IP address from the test system. In addition to changing its connection approach, Honeypot Hunter could change its static strings. Its false mail server could generate a variety of responses, which would make detecting it more difficult. Furthermore, elaborate anti-detection attempts to determine the actual type of system at the end of the proxy connection could give the honeypot away. Finally, the variable header formats that Send-Safe bulk mailer provides could be adapted to Honeypot Hunter; future versions of the tool thus would generate no distinct test emails. Extending anti-honeypot and detection techniques to better suit non spam groups is another logical next step.

10.3.2 Honeypot Detection in Advanced Botnet Attacks

Internet users have been attacked contineously by widespread email viruses and worms However, in recent times (after 2004) major virus or worm outbreak causing great loss has not been seen. This is not because the Internet is much more secure, but more likely because attackers have shifted their attention to compromising and controlling victim computers, an attack scheme which provides more potential for personal profit and attack capability. This lucrative attack theme has produced a large number of botnets in the current Internet. A "botnet" is a network of computers that are compromised and controlled by an attacker [1]. Each compromised computer is installed with a malicious program called a "bot", which actively communicates with other bots in the botnet or with several "bot controllers" to receive commands from the botnet owner, or called "botmaster". Botmasters maintain complete control of their botnets, and can conduct distributed denial-of-service (DDoS) attacks, email spamming, keylogging, abusing online advertisements, spreading new malware, etc.

In recent years, honeypots have become popular, and security researchers have generated many successful honeypot-based attack analysis and detection systems. As more people begin to use honeypots in monitoring and defense systems, botmasters constructing and maintaining botnets will sooner or later try to find ways to avoid honeypot traps. Botmasters might attempt to remove honeypot traps when constructing and maintaining their botnets. Unlike hardware or software specific honeypot detection methods the honeypot detection methodology presented here is based on a general principle that is hardware and software independent: security defenders who set up honeypots have liability constraint such that they cannot allow their honeypots to send out real attacks to cause damage to others, while botmasters do not need to follow this constraint. As laws are developed to combat cybercrime in the coming years, security experts deploying honeypots will probably incur more liability constraint than they have today, because they knowingly allow their honeypots to be compromised by attackers. If they fail to perform due diligence by securing their honeypot from damaging other machines, they will be considered negligent. In addition, botmasters can detect bot controllers hijacked via DNS redirection by checking whether the IP addresses resolved by DNS queries match the real IP addresses of their bot controllers. Compared with the currently popular hierarchical botnets, a P2P botnet is much harder for the security community to monitor and eliminate. Here a simple but effective P2P botnet construction technique via a novel "two-stage reconnaissance" Internet worm attack is presented which is also capable of detecting and removing infected honeypots during the worm propagation stage.

Honeypot Detection in Hierarchical Botnets

Most botnets currently known in the Internet are controlled by botmasters via a hierarchical network structure. Figure 10.1 shows the basic network structure of a typical botnet (for simplicity, only a botnet with two bot controllers shown). All compromised computers in a botnet are called "bots". They frequently attempt to connect with one or several "bot controllers" to retrieve commands from the botnet attacker for further actions. These commands are usually issued from another compromised computer (to hide botmaster's real identity) to all bot controllers. To prevent defenders from shutting down the command and control channel, botmasters usually use multiple redundant bot controllers in their botnets.

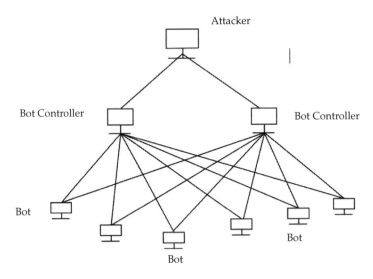

Figure 10.1 Illustration of a hierarchical botnet.

To set up bot controllers flexibly, botmasters usually hard-code bot controllers' domain names rather than their IP addresses in all bots. Botmasters also try to keep their bot controllers mobile by using dynamic DNS (DDNS), a resolution service that facilitates frequent updates and changes in machine location. Each time a bot controller machine is detected and shut down by its user, botmasters can simply create another bot controller on a new compromised machine and update the DDNS entry to point to the new controller.

The above honeypot detection technique will also falsely treat some normal computers as honeypots: these computers are subject to security egress filtering such that their outgoing malicious traffic are blocked. This false positive in honeypot detection, however, does not matter to botmasters. If a bot computer cannot send out malicious traffic, it is better be removed from a botnet, no matter whether it is a honeypot or a well managed normal computer. To detect a hijacked bot controller in a hierarchical botnet, botmasters can issue a test command via the bot controller under inspection that causes botnet members to send trivial traffic to the botmasters' "sensors". The hijacked controller can then easily be detected if the command is not carried out or is not carried out correctly.

Detection of Honeypot Bots

First, we introduce a method to detect honeypots that are infected and acting as bots in a botnet [1]. The general principle is to have an infected computer send out certain malicious or "counterfeit" malicious traffic to one or several remote computers that are actually controlled by the botmaster. These remote computers behave as "sensors" for the botmaster. If the sensors receive the "complete" and "correct" traffic from the infected host, then the host is considered "trusted" and is treated as a normal bot instead of a honeypot. Since honeypot administrators do not know which remote computers contacted are the botmaster's sensors and which ones might be innocent computers, they cannot defend against this honeypot detection technique without incurring the risk of attacking innocent computers.

This honeypot detection procedure is illustrated in Figure 10.2. A newly infected computer cannot join a botnet before it is verified. This potential bot machine must first send out malicious traffic to many targets, including the botmaster's secret sensor (unknown to the newly infected machine). When the botmaster's sensor receives the traffic and verifies the correctness of the traffic (ensuring that

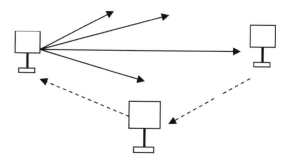

Figure 10.2 Illustration of the procedure in detecting honeypot bots in a hierarchical botnet.

it was not modified by a honeypot), the sensor informs the bot controller of the bot's IP address. The bot controller then authorizes the checked bot so that the bot can join the botnet. To prevent the possibility of a single point of failure, a botmaster could set up multiple sensors for this test.

This honeypot detection procedure can be performed on a newly infected computer before it is allowed to join a botnet. Such a botnet has a built-in authorization mechanism. The botmaster (or the botnet controller) uploads the authorization key to the host and allows it to join the botnet only after the host passes the honeypot detection test. In addition, botmasters may perform the honeypot detection periodically on botnets to discover additional honeypot bots. This could be done whenever botmasters renew their bots' authorization keys or encryption keys, or update the botnet software. This honeypot detection scheme relies on the report of sensors deployed by botmasters. Therefore, botmasters must first ensure that sensor machines themselves are not honeypots. This is not hard to be done since only a few sensor machines are needed-botmasters can manually investigate these machines thoroughly beforehand.

Detection through Infection

When a computer is compromised and a bot program is installed, some bot programs will continuously try to infect other computers in the Internet. In this case, a honeypot must modify or block the outgoing malicious traffic to prevent infecting others. Based on this liability constraint imposed on honeypot security professionals, a botmaster could let compromised computers send malicious infection traffic to her sensors.

Some honeypots, such as the GenII honeynets have Network Intrusion Prevention System (NIPS) that can modify outbound malicious traffic to disable the exploit. To detect such honeypots, botmasters' sensors need to verify that the traffic sent from bots are not altered (e.g., using MD5 signature). It is also important that a newly compromised bot does not send malicious traffic to the sensors alone after the initial compromise. It must hide the honeypot checking procedure to prevent defenders from allowing the initial honeypot detection traffic going out. To hide the sensor's identity, a bot could put the sensors' IP addresses at a random point in the IP address list to be scanned. For a bot that infects via email, the sensors' email addresses could be put at a random point in the outgoing email address list. This procedure will delay the newly infected computer's participation in the botnet, but a botmaster would be willing to incur this slight delay to secure their botnet, because a botnet has long term use to its botmaster.

This honeypot detection technique is difficult for honeypot defenders to deal with. Honeypot defenders cannot block or even modify the outgoing infection traffic. Without accurate binary code analysis, honeypot defenders will not be able to know which target IPs belong to the botmaster's sensors. A botmaster can make the code analysis even harder by obfusticating or encrypting sensors' IP addresses in the code.

Detection through other Illicit Activities

Based on our general honeypot detection principle, botmasters can have their botnets send out other types of illicit traffic to sensors for honeypot detection. These illicit activities include:

Low rate port scanning. By hiding sensors' IP addresses in the port-scan IP address list, a bot can detect whether or not it is in a honeypot that limits outgoing connection requests. For example, GenII honeynet limits the number of outbound connection rate. Some normal computers are configured (e.g., installed a firewall, or a worm detection software such as Kreibich et al. (2005)) to limit outgoing connection rate as well. To avoid mislabeling such computers as honeypots, and also to avoid possible detection by users, botmasters should let their bots conduct a very low rate stealthy port-scan for honeypot detection. Email spamming. A botmaster could also have a bot send out spam email to one or several target email addresses owned by the botmaster. These e-mail addresses behave as the honeypot detection sensors. Outgoing email spam, such as "phishing" email could make honeypot security professionals liable for substantial financial losses if they reach real users.

Detection of Hijacked Bot Controllers

Here techniques to detect hijacked bot controllers are introduced. With the help from Dynamic DNS providers, Dagon et al. presented an effective botnet sinkhole that can change the domain name mapping of a detected bot controller to point to a monitoring machine. This way, the monitor receives connection requests from most (if not all) bots in the botnet. Conceptually speaking, the monitor becomes a hijacked bot controller, which is similar to a honeypot in term of functionality. From a botmaster's perspective, the botnet monitor is very dangerous, because security professionals can learn most of the IP addresses of bots in a botnet — the monitor owners can easily provide a "black-list" of these IP addresses to the security community or potential victims. For this reason, botmasters will do everything they can to eliminate a hijacked bot controller from their botnets. Here two different techniques that botmasters might use to achieve this goal are presented.

Bot Controller DNS Query Check

When a bot controller is hijacked by the DNS redirection method the IP address of the bot controller returned by DNS query will not be the IP address of the real bot controller. Although bots in a botnet know the domain names instead of the actual IP addresses of bot controllers, the botnet owner can easily learn all the IP addresses of the botnet's controllers, because these computers are compromised by the botmaster and are running the botmaster's bot controlling program.

Therefore, a botmaster can keep an up-to-date DNS mapping table of all bot controllers. Using one compromised computer as a sensor, the botmaster can have this sensor continuously send DNS queries to resolve the name and IP mapping of all bot controllers in the botnet and then compare the results with the real domain name mapping table. Besides the short time period right after the botmaster changes the bot controller's IP address, this continuous DNS query procedure is always able to detect whether or not a hijacked bot controller is present in the botnet. If a hijacked controller is detected, the botmaster can immediately use other bot controllers to issue a command to update the domain names in all bots, thus obviating further compromise from the hijacked controller.

Bot Controller Command Channel Check

The above DNS query check is an effective way to detect DNS redirection of bot controllers. However, it is possible for security defenders to conduct a more stealthy monitoring by actually capturing and monitoring a bot controller machine. In this case, the DNS query check will not work. To detect such a physically hijacked bot controller, a botmaster can use the same honeypot detection principle as above. The botnet owner checks whether or not a bot controller passes the botmaster's commands to bots. The monitor presented in Dagon et al. (2006) is called "sinkhole" because it does not pass any botmaster's commands to bots. In fact, a hijacked bot controller puts a much more serious liability burden on security defenders than a normal compromised honeypot. If it passes a botmaster's command to bots in a botnet, the defender could potentially be liable for attacks sent out by thousands of computers in the botnet. For this reason, security defenders do not dare to let a hijacked bot controller send out a single command. Even if the command seems harmless from previous experience, it is always possible that a botnet implements its unique command system. In this case, a known trivial command based on previous experience could possibly conduct harmful task such as deletes files on all of the compromised computers, or launches DDoS attacks against risky targets. Based on this, a botmaster can issue a trivial command to the bot controller under inspection without passing the command to other bot controllers. The trivial command orders a small number of bots to send a specific service request to the botmaster's sensor (e.g., a compromised web server). Bots will not be exposed by this action since they simply send out some normal service requests. If the sensor does not receive the corresponding service requests, the botmaster knows that the bot controller has been hijacked (or is at least not working as required).

Honeypot Detection in P2P-Structured Botnets

To botmasters, changing a botnet's control architecture to be peer-to-peer is a natural way to make a botnet harder to be shut down by defenders. In recent years, botmasters have tested and implemented different kinds of preliminary P2P botnets such as Slapper, Sinit, Phatbotand Nugache. Some researchers have studied P2P botnet designs. Therefore, more P2P botnets will be created in the near future. Botmasters will need to come up with a new honeypot detection technique for a P2P botnet. In a P2P botnet, each bot contains a list of IP addresses of other bots that it can connect with, which is called "peer list". Because there are no centralized bot controllers to provide authentication in a P2P botnet, each bot must make its own decision, or collaborate with its peers, to decide whether its hosted machine is a honeypot or not.

Two-stage Reconnaissance Worm

A two-stage reconnaissance worm is designed to have two parts: the first part compromises a vulnerable computer and then decides whether this newly infected machine is a honeypot or not; the second part contains the major payload and also the authorization component allowing the infected host to join the constructed P2P botnet. Due to the different roles in a worm propagation, the first part is called the "spearhead", and the second part the "main-force" of the worm.

A simple way to verify whether a newly compromised host is a honeypot or not is to check whether or not the worm on it can infect other hosts in the Internet. Figure 10.3. illustrates the propagation procedure of a two-stage reconnaissance worm in infecting host B and checking whether it is a honeypot or not. First, the vulnerable host B is infected by the spearhead of the worm, which contains the exploiting code and the peer list. Second, the spearhead on host B keeps scanning the Internet to find targets (such as host C) to infect with the spearhead code. Third, after the spearhead on host B successfully compromises m hosts (include both vulnerable and already-infected ones), it tries to download the main-force of the worm from any host in its peer list that has the main-force component. The main-force code lets the worm join the constructed botnet via the authorization key contained in the main-force (e.g., the authorization key could be a private public key). By deploying such a two-stage reconnaissance worm, the botnet is constructed

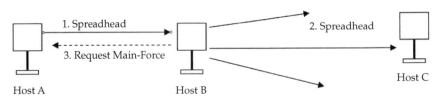

Figure 10.3 Illustration of the propagation procedure of a two-stage reconnaissance worm.

with a certain time delay as the worm spreads. This means that some infected hosts will not be able to join the botnet, since they could be cleaned before the main-force is downloaded. However, this does not affect the botnet, since it makes no difference to the botmaster whether or not the botnet contains bots that will be quickly removed by security defenders. In fact, it is not a new idea to spread a worm in two stages. Blaster worm and Sasser worm used a basic FTP service to transfer the main code of the worm after compromising a remote vulnerable host (CERT). The two-stage reconnaissance worm presented here can be treated as an advanced two-stage worm by adding the honeypot detection functionality into the first-stage exploit code.

The reconnaissance worm described above needs a separate procedure to obtain the complete bot code. This could be a problem for a botnet since the original Host A might be unaccessible from others, or Host A has changed its IP address when Host Be tries to get the main-force worm code. To deal with this issue, the worm could combine the main-force code together with the spearhead code, but first deactivate and possibly encrypt the main-force code at the beginning. After the spearhead code verifies that a hosted machine is not honeypot, it will unpack and execute the main-force code. One drawback of this approach is that honeypot defenders can easily obtain the main-force code even when their honeypots are not able to join the botnet.

10.3.3 Mapping Internet Sensors With Probe Response Attacks

The growing use of Internet sensors as a tool to monitor Internet traffic allows one to determine the location of sensors using the networks that publicly report statistics. In particular, a new "probe response" attack technique is proposed with a number of optimizations for locating the sensors in currently deployed Internet sensor networks and illustrate the technique for a specific case study that shows how the attack would locate the sensors of the SANS Internet Storm Center (ISC) using the published data from those Sensors analysis center can be used to mount an attack. A new class of attacks capable of locating Internet sensors that publicly display statistics will be seen. This gives insights into the factors which affect the success of probe response attacks. Also countermeasures that protect the integrity of Internet sensors will be discussed and still allow for an open approach to data sharing and analysis. Without public statistics, the benefits of a widely distributed network of sensors are not fully realized as only a small set of people can utilize the generated statistics.

A detailed algorithm which uses a straightforward divide and conquer strategy along with some less obvious practical improvements to map the sensor locations using information found in the ISC port reports is presented.

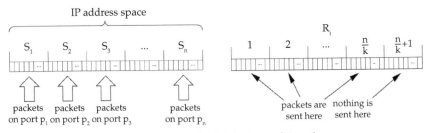

Figure 10.4 The First (left) and Second (right) Stage of Attack.

The core idea of the attack is to probe an IP address with activity that will be reported to the ISC if the addresses are among those monitored, then check the reports published by the network to see if the activity is reported. If the activity is reported, the host probed is submitting logs to the ISC. Since the majority of the reports indicate an attempt to make a TCP connection to a blocked port (which is assumed to be part of a search for a vulnerable service), a single TCP packet will be detected as malicious activity by the sensor. To distinguish our probe from other activity on that port, enough packets need to be sent to significantly increase the activity reported. As it turns out, a number of ports normally have little activity, so this is not burdensome. This probing procedure is then used for every possible IP address. It is quite possible to send several TCP/IP packets to every address. The simplest way to find all hosts submitting logs to the ISC is then to send packets to the first IP address, check the reports to determine if that address is monitored, send packets to the second IP address, check the reports again, and so on.

However, some time must be allowed between sending the packets and checking the reports. Participants in the ISC network typically submit logs every hour, and additional time should be allowed in case some participants take a little longer, perhaps for a total wait of two hours. Obviously, at this rate it will take far too long to check every IP address one by one. In order for a sensor probing attack to be feasible, many addresses need to be tested at the same time.

Two observations will help us accomplish this. First, the vast majority of IP addresses either do not correspond to any host, or correspond to one that is not submitting logs. With relatively few monitored addresses, there will necessarily be large gaps of unmonitored address space. Hence, it will not be possible to rule out large numbers of addresses at a time by sending packets to each, then checking if any activity is reported at all. If no activity is reported, none of the addresses are monitored. Sending packets to blocks of addresses numerically adjacent is likely to be especially effective, since monitored addresses are likely to be clustered to some extent, leaving gaps of addresses that may be ruled out. Second, since malicious activity is reported by port, we can use different ports to conduct a number of tests simultaneously. These considerations led the authors to the method described in the following section. It is worth noting that the problem solved by this algorithm is very similar to the problems of group blood testing. However, much of theoretical results from this area focus on optimizing the solutions in a different way than we would like to and thus are not directly applicable to this problem.

10.4 COUNTERMEASURE FOR DETECTION OF HONEYPOT DEPLOYMENT

The honeypot-aware botnet introduced in above section relies on the basic principle that security professionals have liability constraints, while attackers do not need to obey such constraints. The fundamental counterattack by security professionals, therefore, is to invalidate this principle [2]. Of course, the law currently regulating cyberspace security is not mature or defined in many countries; hence, some researchers or security defenders have deployed honeypots that freely send out malicious attacks.

However, such honeypot defense practices are negligent and unethical. It will become illegal as the laws regarding cyberspace security and liability gradually mature. The current popular GenII honeynet has considered preventing attack traffic from being sent, but it does not implement this strictly. First, it limits outgoing connection rate, thus it is possible that some early honeypot detection traffic could be sent out. Second, it can block or modify only detected outgoing malicious traffic, thus unknown malicious packets are possibly being sent out by honeypots. For this reason, the GenII honeynet might be able to avoid the honeypot detection conducted by attackers; but at the same time, it could actually infect other computers as well and thus potentially make the honeynet owners liable for the ensuing damage.

When botmasters deploy sensors to detect honeypots by checking test traffic, they rely on the fact that the identities of sensor machines are secret to honeypot defenders. Therefore, if security defenders could quickly figure out the identities of sensors before botmasters change their sensor machines (such as through binary code analysis), defenders' honeypots could avoid detection by allowing test traffic going out to those sensors.

A promising defense against honeypot-aware attacks is the "double-honeypot" idea. It can be seen here that attackers need to take complicated extra steps in order to avoid being fooled by double-honeypot traps. By using dual honeypots, or a distributed honeypot network that can accurately emulate the network traffic coming in from the Internet, security defenders can take proactive roles in deceiving honeypot-aware attacks. For example, security defenders can build a large-scale distributed honeynet to cover many blocks of IP space, and allow all malicious traffic to pass freely within this honeypot virtual network. However, this defense will be ineffective if attackers use their own sensors to detect honeypots.

10.4.1 The Honeyanole System

A deceptive system, called honeyanole [3], is proposed to escape from honeypot hunters as well as to collect attacking information to enhance further defense. In this system, non-attacking service connections and probing connections are monitored and transmitted, while the attacking service connections are transparently redirected to the fabricated system for the attacking process collection. A common countermeasure against the deployment exposure is to redirect the connection to avoid directly interacting with a honeypot. The redirection technique is to decompose Internet traffic into two destinations: a production server or a honeypot.

In honeyanole, both layer-2 and layer-3 redirection mechanisms are employed to dynamically transmit incoming traffic flows for the purpose of resisting the detection of honeypot hunters as well as collecting and analyzing attacking information. The network connections are categorized into regular service requests, probe requests, and attacking service requests. Under the layer-2 redirection, regular service connections and probe requests are directed to the real system. In this case, the redirection latency is insignificant, and hence the honeypot is not suspicious to honeypot hunters. Once an attacking service connection is discovered, layer-3 redirection is active and the connection is redirected to the fabricated system.

There are three phases in honeyanole:

1. collection phase,
2. redirection phase,
3. and deception phase.

The main task of collection phase is to build a blacklist of possible attackers to support the redirection server. As shown in Figure 10.5, all traffic flows from Internet to production server will be mirrored to the detection module for intrusion inspecting. The information of possible attackers will be gathered by collection module from detection module and other three defensive systems, including the illegal access log, the record of probes, and exchanged defensive information.

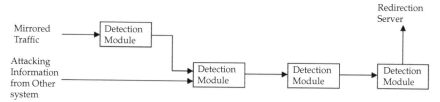

Figure 10.5 Modules inside the collection phase.

After the collection, the alerts of attacking information for eliminating the same attack and incurring a new threat based on alert type, source address, and target address are raised. Then, the analysis module performs the correlation of collected attacking information to predefine attack scenarios, such as network scans, port scans, or vulnerability attacks. Upon finishing the analysis, the decision module would build an orderly list of possible attackers according to temporal information and involved services. Finally, a blacklist is distributed to redirection server dynamically. For redirection, the server with external, internal, and redirection interfaces are designated to connect to Internet, a production network, and a deception server respectively. When an incoming traffic arrives from Internet interface, redirection module will transmit it to a production server or a deception server with the aid of the blacklist. Operational flows of the redirection module can be depicted in Figure 10.6.

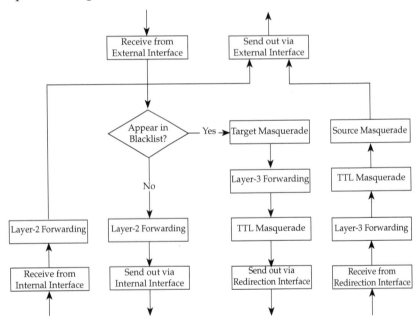

Figure 10.6 Operation flows of redirection module.

If an incoming traffic is a probe or its source address does not appear in the blacklist, the layer-2 redirection would forward the traffic to a production server via the internal interface without changing any packet's content. However, if an incoming traffic has its source address appearing in the blacklist, the layer- 3 redirection will take place. With layer-3 redirection, target masquerade changes the target address of incoming packets into the deception server before layer-3 forwarding. And, TTL masquerade adjusts the value of ttl in IP header to conceal from the action of layer-3 forwarding. Similarly, the outgoing packets will be adjusted accordingly.The deception phase is responsible to capture the intrusive processes.

With honeyanole, various types of honeypots systems can be deployed as deception servers. Adopting a high-interaction honeypot can obtain more intrusive information and easily suffer from deployment disclosure by system level detections, while a low-interaction honeypot could be discovered by service support detections. How to precisely predict an intrusion, more specifically honeypot detection, is the key feature to deception detection. By combining the above three phases, honeyanole system is built as shown in Figure 10.7.

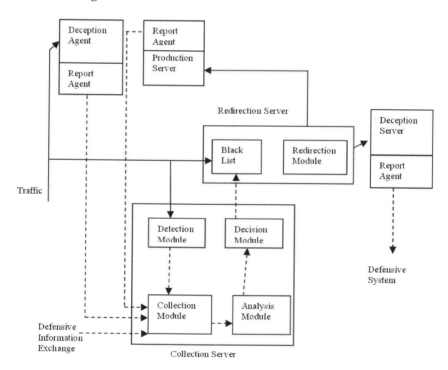

Figure 10.7 The global view of honeyanole.

Therefore, how to build an effective and accurate blacklist is an imperative task. All traffic to production server, including service connections and attacks, are mirrored to detection module to execute an intrusive inspection. The alert generated by the detection module is also the main part of attacking information.

10.4.2 A Hybrid Honeypot Architecture for Scalable Network Monitoring

Here a globally distributed, hybrid monitoring architecture is proposed which provides scalable, early warning and analysis of new Internet threats like worms or automated attacks. It can capture and analyze new vulnerabilities and exploits as they occur. The architecture increases the exposure of high-interaction honeypots to these threats by employing low-interaction honeypots as frontend content filters. Host-based techniques capture relevant details such as packet payload of attacks while network monitoring provides wide coverage for quick detection and assessment. Also for reducing the load of the backends, it filters prevalent content at the network frontends and uses a novel handoff mechanism to enable interactions between network and host components.

Lightweight virtual honeypots can instrument a large address space effectively. On the other hand, they do not support full behavioral fidelity, e.g., a new threat may fail to be accurately captured by a low-interaction honeypot. This is almost the reverse for high-interaction honeypots. A new threat can successfully interact with such a machine, but the high-interaction system does not have optimal performance and would not scale to a large address space. This advantage of the scalability provided by the low-interaction virtual honeypots while still being able to provide detailed behavioral analysis is considered here. This is achieved by combining the scalability and interactivity of different honeypot technologies through a hybrid system as shown in Figure 10.8.

This hybrid architecture consists of three components:

1. Lightweight Honeypots. Scalable low-interaction honeypots are widely deployed and monitor numerous variable sized unused network blocks. These boxes filter requests to a set of centralized, high-interaction honeypots.

2. High Interaction Honeypots. A collection of centralized VMware (virtual machine) hosts running a wide variety of host operating systems and applications. Filtered traffic from the lightweight sensors are directed here for behavioral analysis.

3. Command and Control. A centralized mechanism provides a unified view of the collection infrastructure including overall traffic rates, source and destination distributions, as well as global payload cache state. This mechanism coordinates connection redirection across sensors. It also monitors backend performance and provides feedback to the frontend redirection mechanisms.

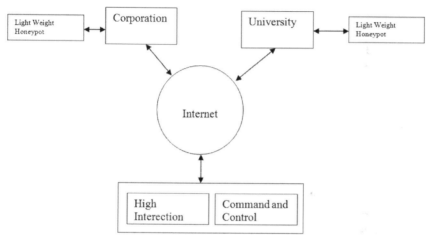

Figure 10.8 A Hybrid Honeypot Architecture.

All of our address space is instrumented by low-interaction honeypots. The role of these honeypots is to filter the uninteresting traffic by answering it locally. Interesting traffic is forwarded to a set of backends that constitutes our high-interaction honeypots. One classify uninteresting traffic roughly as all packets that do not belong to an existing connection, SYN packets that do not lead to an established three-way handshake, payloads that we have seen many times before. To monitor for the occurrence of new threats that have the same first payload, sampled connection handoff are supported, periodically sending known signatures to the backend. This sampling is dynamic relying on a feedback mechanism from the profiled backend honeypots.

The backends run VMware to provide several high-interaction systems per physical machine. Their network setup prevents them from contacting the outside world. However, instead of blocking outgoing connections, we redirect them back to the heavyweight honeypots. If a worm was to infect one of the machines, it would successfully infect more susceptible backends as long as uninfected backends are available. This mechanism is unique in that it allows to watch actual worm propagation, including exploit as well as payload delivery in a controlled environment. Once this type of propagation is detected, the checkpointing features of VMware will not only allow to save these infected states and corresponding forensic information, but also return the machines to a known good state for further use in analysis.

The final piece of the architecture is a control component. The control component aggregates traffic statistics from the lower-interaction frontends and monitors the load and other data from the backends. The control component is responsible for analyzing all received data for abnormal behavior such as a new worm propagating. This is achieved by two separate analysis phases that we combine at the end. One phase analyzes the network data of the frontends for statistics that relate to worm propagation: e.g., increases in source IP addresses for a certain port, increases in scanned destination IP addresses for a certain port, and all of this correlated with payload data. The other phase keeps track of the behavior of the backends. If noticed, the backends initiate unusual network connections that is normally a sign of worm infection. These unusual network connections and other forensic information can be stored when abnormal behavior has been detected.

10.5 SUMMARY

It is a difficult problem to deploy honeypots that cannot be detected by hackers. We must remember that honeypot technology is only effective if an attacker does not know she is attacking a "trap" instead of a real system. Therefore, it is critically important for security professionals who deploy honeypots to be aware of the methods blackhat hackers use to identify them.

The attack to a honeypot can be carried out any level as discussed in Network issues and System issues. There are lots of techniques present for fingerprinting the presence of honeypot in the network. Various techniques for honeypot detection is also discussed along with some of the honeypot hunter which shows that potentially all the types of honeypot can be detected at some or the other time. At last various works are proposed for countermeasures for honeypot detection. Thus both side the attacker and defender of the honeypot is now ready with their armor and weapons and in recent times there may be more powerful internet tools and techniques may be invented in both the sides.

EXERCISES

1. "The arms race between whitehats and blackhats has begun", Explain.
2. Outline the network fingerprints that detect the presence of honeypots.
3. What are the system blueprints that detect the presence of honeypots?
4. Elaborate the working of Honeypot Hunter.
5. Explain the working of Honeyanole System.
6. What are the various ways for detection of honeypot deployement?

REFERENCES

[1] P. Wang, L. Wu, R. Cunningham, and C.C. Zou, "Honeypot detection in advanced botnet attacks," *International Journal of Information and Computer Security*, vol. 4, pp. 30–51, 2010.

[2] L.M. Shiue and S.J. Kao, "Countermeasure for detection of honeypot deployment," in *International Conference on Computer and Communication Engineering, ICCCE 2008*, Kuala Lumpur, 2008, pp. 595–599.

[3] J. Bethencourt, J. Franklin, and M. Vernon, "Mapping internet sensors with probe response attacks," in *Proceedings of the 14th conference on USENIX Security Symposium*, vol.14, Baltimore, MD USA, 2005, pp. 193–208.

11

Honeypots and Network Forensics

In network security, an attack is associated with pre-event (before the attack), during-event (during the attack), and post-event (after the attack) scenarios. In earlier chapters Honeypots were shown playing a critical role against attacks in pre-event and during event scenarios. This chapter elaborates the role of honeypots in post event scenarios. Network forensics is the science that deals with detection and investigation of intrusions in the post-event scenarios. In the starting a background of network forensics is presented. Then, Honeypots as a network forensic analysis tool (NFATs) is discussed. The chapter explains the general process model of forensics and describes the role of Honeypot in its different phases. Next, it outlines the Honeypot based Network Forensics Frameworks. Finally the major challenges in this upcoming and young discipline are presented.

Network forensics is the science that deals with capture, recording, and analysis of network traffic. Honeypots play an important role for forensics and investigation of networks. The network log data is collected from existing security products like Kismet, Wireshark, etc., analyzed for attack characterization and investigated to traceback the perpetuator. Network forensics is not another term for network security. Network forensics can be considered as an essential part in Network security. Earlier, the data for forensic analysis was collected from security products like firewalls and intrusion detection systems only. With their evolution, Honeypots have become key contributor in capturing the attack data which is analyzed and investigated, hence facilitating the process of network forensics. Network forensics, however, may involve certain crimes which are legally prosecutable but which may not breach network security policies [1].

Case Study

On August 6, 2009, social networking sites like Twitter, Facebook and Google blogger were knocked down by Distributed Denial of Service (DDoS) attacks. Facebook and Google could recover within a day while Twitter staff team worked round the clock in the weekend to deal with the attack. Los Angeles Times speculated that the perpetrators of the DDoS attack may have been bored teenagers or Russian and Georgian political operatives involved in cyberspace fighting. The newspaper quoted security experts that fingerprints of a sophisticated operation involving botnets was observed and the Twitter website had limited capacity to handle incoming traffic which caused the service to become unavailable.

The obvious reason for the success of this attack was that Twitter's network did not have the defenses in place to mitigate a massive DDoS attack. Most traditional security products aren't equipped to handle the massive bombardment of packets that happens in a DDoS attack. The lack of solid contingency plan and pro-active security mechanism created a fragile platform vulnerable to attack.

Having the appropriate tools in place and following the correct procedures would have helped eliminate or mitigate the effects of DDoS attack. A network analysis tool can be used to capture all packets in a common data format for analysis. It can also raise alerts when thresholds are exceeded. Network forensics tools can be used to reconstruct the sequence of events that occurred at the time of the attack which will give a complete picture. If the present attack could not be prevented, crucial information is gained to prevent a similar attack in the future. Network forensics can be used to analyze how the attack occurred, who was involved in that attack, the duration of the exploit, and the methodology used for the network attack. It also helps in characterizing zero day attacks. In addition, network forensics can be used as a tool for monitoring user activity, business transaction analysis and pinpointing the source of intermittent performance issues.

Network security protects the system against attack while network forensics does not. Network security products are generalized and look for possible harmful behaviors and they monitor the network 24 hours a day. Network forensics is post mortem investigation of the attack. It is also case specific as each crime scenario may be different and the process is initiated *notitia criminis* (after crime notification).

Network forensics is a natural extension of computer forensics. Computer forensics was introduced by law enforcement and has many guiding principles from the investigative methodology of judicial system. Computer forensics involves preservation, identification, extraction, documentation, and interpretation of computer data. Network forensics evolved as a response to the hacker community and involves the capture, recording, and analysis of network events in order to discover the source of security attacks.

In computer forensics, the investigator and the person being investigated are on two different levels with the investigator at an advantage. In network forensics, the network investigator and the attacker are at the same skill level. The network forensic specialist uses many of the same tools and engages in the same set of practices as the person being investigated [2]. The difference in them is based on ethical standards and not on technical skills.

Network forensics involves monitoring network traffic and determining if there is an anomaly in the traffic and ascertaining whether it indicates an attack. If it is so then the nature of the attack is also determined. When attacks are successful, forensic techniques enable investigators to catch the attackers. The ultimate goal is to provide sufficient evidence to allow the perpetrator to be prosecuted.

11.1 NETWORK FORENSICS

The concept of network forensics deals with the data found across a network connection, mostly ingress and egress traffic from one host to another. Network forensics tries to analyze the traffic data logged through firewalls or intrusion detection systems or at network devices like routers and switches as shown in Figure 11.1.

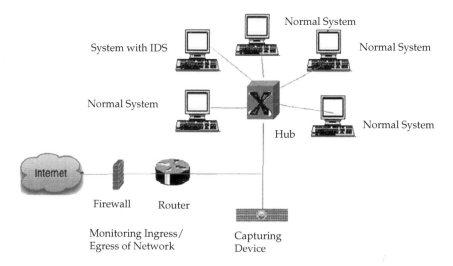

Figure 11.1 Network with traffic monitoring tools.

Network forensics is defined in [3] as "the use of scientifically proven techniques to collect, fuse, identify, examine, correlate, analyze, and document digital evidence from multiple, actively processing and transmitting digital sources for the purpose of uncovering facts related to the planned intent, or measured success of unauthorized activities meant to disrupt, corrupt, and or compromise system components as well as providing information to assist in response to or recovery from these activities."

Marcus Ranum [4] is credited with coining the word *network forensics* as "the capture, recording, and analysis of network events in order to discover the source of security attacks or other problem incidents."

11.1.1 Classification of Network Forensics Systems

Network forensic systems are classified into two types each based on various characteristics like purpose, capture, and platform, time of analysis and data source. Figure 11.2 shows give the summary of classification

On the basis of purpose, Network Forensics Systems can be classified as 'Generic Network Forensics' and 'Strict Network Forensics'. *General Network Forensics (GNF)* focuses on enhancing security [5]. The network traffic data is analyzed and some patterns of the attacks are discovered. *Strict Network Forensics (SNF)* involves rigid legal requirements as the results obtained will be used as evidence for prosecution of the network crimes.

On the basis of Packet Capture, Network Forensics Systems can be classified as *Catch-it-as-you-can systems* and *Stop, look and listen systems*. *'Catch-it-as-you-can'* systems are those where all packets passing through a particular traffic point are captured and stored. *'Stop-look-and-listen'* systems analyze eack packet in memory and only some information is saved for future.

Platform classifies network forensic systems into *hardware appliance and pre-installed software*. The *hardware appliance* can capture data, analyze and present the results on a computer interface. The system can be exclusively *software also,* which will be installed on a host and analyze stored packet captures or net flow records.

Network forensic systems can also be classified on the basis of Time of Analysis. Commercial network forensic analysis appliances involve *real time* network security surveillance, signature-based anomaly detection, data analysis and forensic investigation. Many open source software tools are designed for *post mortem* investigation of packet captures [6]. Full packet captures are performed by sniffer tools, data is stored in a host and analysis is performed off line at a later time.

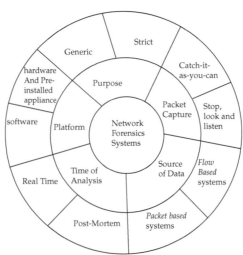

Figure 11.2 Classifications of Network Forensics Systems.

Data Source divides the network forensic systems into *Flow based* systems and *Packet based* systems. *Flow based* systems collect statistical information based on some criteria within the network traffic as it passes through the network. The network equipment collects this data and sends it to a flow collector which stores the data. *Packet based* systems involve full packet captures at various points in the network. The packets are collected and stored for deep packet inspection.

11.1.2 Motivation for Network Forensics

There are a number of forces driving network forensics [7] and some of them are:

- The increase in the number of security incidents affecting many organizations and individuals and the increase in the sophistication of the cyber attacks
- The attacker is covering the tracks used to cause the attacks, making it more difficult to traceback.
- Companies doing business on the Internet cannot hide a security breach and are now expected to prove the state of their security as a compliance measure for regulatory purposes
- Internet Service Providers (ISPs) are also being made responsible for what passes over their network
- The defensive approaches of network security like firewalls and intrusion detection systems can address attacks only from the prevention, detection and reaction perspectives.

The alternate approach of network forensics becomes important as it has the investigative features. Network forensics ensures that the attacker spends more time and energy to cover his tracks making the attack costly. Network criminals are also cautious to avoid prosecution for their illegal actions. This acts as a deterrent and reduces network crime rate thus improving security.

The ISO 27001 / 27002 standard [8, 9] (Information technology—security techniques—information security management) specifies the requirements for establishing, implementing, operating, monitoring, reviewing, maintaining and improving a documented Information Security Management System (ISMS) within the context of the organization's overall business risks. Comprehensive audit data is to be maintained to meet the compliance requirements of regulations like Sarbanes-Oxley (SOX) Act, Gramm-Leach-Bliley Act (GLBA), Health Insurance Portability and Accountability Act (HIPAA), Federal Information Security Management Act (FISMA), and Payment Card Industry (PCI) Data Security Standard (DSS). The integration of these monitoring policies into infrastructures will facilitate the verification of compliance. Hence having the network forensics process in place will meet all the requirements as mentioned above.

11.1.3 Honeypot Approaches for Network Forensics

There are two ways of developing a network forensic process. One way is to reactively use traditional security products like firewalls & intrusion detection systems, analyze the data and investigate. The other way is to proactively lure the attacker by means of honeypots [10] and honeynets [11] and observe the attack patterns. The behavioral profiles of attackers are created and their exploitation mechanisms are understood.

Since, a Honeynet (or high interaction honeypots) is a highly controlled network of computers, involving real operating systems and applications, designed in a way to capture all activity when attacked so full extent of the attackers' behavior can be learnt by letting these high-interaction honeypots to interact with them [12]. The Honeynet controls the attacker's activity by using a honeywall gateway allowing inbound traffic to the victim systems and controlling the outbound traffic using intrusion prevention technologies.

Virtual honeynet is another solution that allows us to run multiple platforms needed on a single computer. The term virtual is used because all the different operating systems have the 'appearance' to be running on their own, independent computer. The virtualization software allows running multiple operating systems at the same time, on the same hardware. The advantages of virtual honeynets are cost reduction and easier management, as everything is combined on a single system.

In Network Forensics, Virtual honeynets can be used to gather data about the intruder's attack strategy and study the applications used to carry out the attack [13]. This approach gives an edge to network forensic investigator over the attacker, as he is aware of the attack methodology and the tools being used. The investigator is able to fingerprint the attacker with this knowledge.

11.2 HONEYPOT AS NETWORK FORENSIC ANALYSIS TOOLS

Network Forensic Analysis Tools (NFATs) allow administrators to monitor the networks, gather all information about anomalous traffic, assist in network crime investigation and help in generating a suitable incident response. NFATs also help in the following functions [14]:

- insider theft and misuse of resources
- prediction of future attack targets
- protect intellectual property
- perform risk assessments
- exploit (break-in) attempt detection
- data aggregation from multiple sources
- network traffic recording and analysis
- determination of hardware and network protocols in use

NFATs capture the entire network traffic, allow the users to analyze the network traffic according to their needs and discover significant features about the traffic. NFATs synergize with IDSs and Firewalls and make possible the long term preservation of network traffic records for quick analysis [15].

The attack traffic can be replayed and attackers' moves can be analyzed for malicious intent. NFATs facilitate organization of the captured network traffic packets to be viewed as individual transport layer connections between machines. This enables the analysis of protocol layers and packet content. Traffic patterns are extracted between various machines. A description of a partial list of the NFATs [16] is given below:

Description of Network Forensic Analysis Tools

- **NetIntercept**: Captures network traffic and stores in pcap format, reassembles the individual data streams, analyzes them by parsing to recognize the protocol and detect spoofing and generates a variety of reports from the results.

- **NetDetector**: Captures intrusions, integrates signature-based anomaly detection, reconstructs application sessions and performs multi time-scale analysis on diverse applications and protocols. It has an intuitive management console and full standards based reporting tools. It imports and exports data in a variety of formats.

- **NetWitness**: Captures all network traffic, reconstructs the network sessions to the application layer for automated alerting, monitoring, interactive analysis and review.

- **NetworkMiner**: Network traffic capture by live sniffing, performs host discovery, reassembles transferred files, identifying rogue hosts and assesses how much data leakage was affected by an attacker.

- **SilentRunner**: Captures, analyzes and visualizes network activity by uncovering break-in attempts, abnormal usage, misuse and anomalies. It generates an interactive graphical representation of the series of events and correlates actual network traffic. It also plays back and reconstructs security incidents in their exact sequence.

- **Infinistream**: Utilizes intelligent Deep Packet Capture (iDPC) technology and performs real time/back-in-time analysis. It does high speed capture of rich packet details, statistical analysis of packet/flow based data and recognizes hundreds of applications. It uses sophisticated indexing and Smart Recording and Data Mining (SRDM) for optimization.

- **Iris**: Collects network traffic and reassembles it as its native session based format, reconstructs the actual text of the session, replays traffic for audit trial of suspicious activity, provides a variety of statistical measurements and has advanced search and filtering mechanism for quick identification of data.

- **OmniPeek Suite**: Provides real-time visibility into every part of the network. It has high capture capabilities, centralized console, distributed engines, and expert analysis. Omnipliance is a network recording appliance with a multi-terabyte disk farm and high-speed capture interfaces. OmniEngine software running on the device captures and stores network traffic and OmniPeek interface searches and mines the captured data for specific information.

- **Xplico**: Captures internet traffic, dissects the data at the protocol level, reconstructs and normalizes it for use in manipulators. The manipulators transcode, correlate and aggregate it for analysis and presents the results in a visualized form.
- **Solera DS 5150 with DeepSee Suite**: DS 5150 is an appliance for high speed data capture, complete indexed record of network traffic, filtering, regeneration and playback. DeepSee forensic suite has three softwares—Reports, Sonar and Search—to index, search and reconstruct all network traffic.
- **PyFlag**: Python Forensic Log Analysis GUI is an advanced forensic tool to analyze network captures in libpcap format while supporting a number of network protocols. It has the ability to recursively examine data at multiple levels and is ideally suited for network protocols which are typically layered. PyFlag parses the pcap files, extracts the packets and dissects them at low level protocols (IP, TCP or UDP). Related packets are collected into streams using reassembler. These streams are then dissected with higher level protocol dissectors (HTTP, IRC, etc.).
- **Refer [16] for more information about these tools.**

There are many other open source network security and monitoring tools which help in specific activities. These tools were designed with information security in mind rather than evidence processing and hence do not have a forensic standing. A description about a partial list of network security tools is given below [15]:

Description of Network Security and Monitoring Tools

- **TCPDump**: A common packet sniffer and analyzer, runs in command line, intercepts and displays packets being transmitted over a network. It captures, displays, and stores all forms of network traffic in a variety of output formats. It will print packet data like timestamp, protocol, source and destination hosts and ports, flags, options, and sequence numbers.
- **TCPFlow**: Captures data transmitted as part of TCP connections (flows) and stores the data for protocol analysis. It reconstructs the actual data streams and stores in a separate file. TCPFlow understands sequence numbers and will correctly reconstruct data streams regardless of retransmissions or out-of-order delivery.

Description of Network Security and Monitoring Tools (... *contd.***)**

- **TCPStat**: Reports network interface statistics like bandwidth, number of packets, packets per second, average packet size, standard deviation of packet size and interface load by monitoring an interface or reading from libpcap file.

- **TCPReplay**: Suite of tools with the ability to classify previously captured traffic as client or server, rewrite Layer 2, 3 and 4 headers and finally replay the traffic back onto the network. TCPPrep is a multi-pass pcap file pre-processor which determines packets as client or server, TCPRewrite is the pcap file editor which rewrites packet headers, TCPReplay replays pcap files at arbitrary speeds onto the network and TCPBridge bridges two network segments.

- **IOS NetFlow**: Collects and measures IP packet attributes of each packet forwarded through routers or switches, groups similar packets into a flow, to help understand who, what, when, where and how the traffic is flowing. It also detects network anomalies and vulnerabilities.

- **Flow-tools**: Library to collect, send, process and generate reports from NetFlow data. Few important tools in the suite are—flow-capture which collects and stores exported flows from a router, flow-cat concatenates flow files, flow-report generates reports for NetFlow datasets, and flow-filter filters flows based on export fields.

- **NMap:** Utility for network exploration and security auditing. It supports many types of port scans and can be used as on OS fingerprinting tool. It uses raw IP packets in novel ways to determine hosts available on the network, services being offered, operating systems running, firewalls in use and many other characteristics.

- **Ngrep:** A pcap-aware tool that allows specifying extended regular or hexadecimal expressions to match against data payloads. It can debug plaintext protocol interactions to identify and analyze anomalous network communications and to store, read and reprocess pcap dump files while looking for specific data patterns.

- **Ntop:** Used for traffic measurement, traffic monitoring, network optimization & planning, and detection of network security violations. It provides support for both tracking ongoing attacks and identifying potential security holes including IP spoofing, network cards in promiscuous mode, denial of service attacks, trojan horses and port scan attacks.

> ## Description of Network Security and Monitoring Tools (... *contd.*)
>
> - **Wireshark**: Most popular network protocol analyzer. It can perform live capture in libpcap format, inspect and dissect hundreds of protocols, do offline analysis, and work on multiple platforms. It can read and write files in different file formats of other tools.
> - **Snort:** Network intrusion prevention / detection system capable of performing packet logging, sniffing and real-time traffic analysis. It can perform protocol analysis, content searching & matching and application level analysis.
> - **Bro:** A network IDS which detects intrusions by parsing network traffic. It extracts its application-level semantics and executes event-oriented analyzers to compare the activity with patterns deemed troublesome. It is primarily a research platform for traffic analysis and network forensics.
> - **Argus:** Processes packets in capture files or live data and generates detailed status reports of the 'flows' detected in the packet stream. The flow reports capture much of the semantics of every flow with a great deal of data reduction. The audit data is good for network forensics, non-repudiation, detecting very slow scans, and supporting zero day events.
> - **P0f:** Passive OS fingerprinting by capturing traffic coming from a host to the network. It can also detect the presence of firewall, use of NAT, existence of a load balancer setup, the distance to the remote system and its uptime.
> - **SiLK:** System internet Level Knowledge supports efficient capture, storage and analysis of network flow data based on Cisco NetFlow. The tool suite, consisting of collection and analysis tools, provides analysts with the means to understand, query, and summarize both recent and historical traffic data in network flow records. SiLK supports network forensics in identifying artifacts of intrusions, vulnerability exploits, worm behavior, etc. SiLK has performance as a key element and manages the large volume of traffic by storing only the security related information and splits files into predefined categories to reduce lookup time.
> - **Refer [16] for more information about these tools.**

Sebek: A Honeypot Based Tool

Honeynet Project's experiments with kernel-based rootkits for the purpose of capturing the data of interest from within the honeypot's kernel lead to the development of Sebek [17]. This is a data capture tool that lives entirely in kernel space and records all data accessed by users on the honeypot system. It provides capabilities to record keystrokes of a session that uses encryption, recover files copied with SCP, capture and recover passwords used to log in to remote system and enable protected binaries.

Sebek helps in determining when an intruder broke in, how he did it, and what he did after gaining access. It is particularly useful in circumventing encryption by capturing the data post decryption. The idea is to let the standard mechanisms do the decryption work, and then gain access to this unprotected data. The most robust capture method involved accessing the data from within the OS's kernel and there is ample opportunity to improve the subtlety of the technique by hiding the honeypot's actions from all users.

The Architecture of Sebek has two components—client and server—as shown in Figure 11.3:

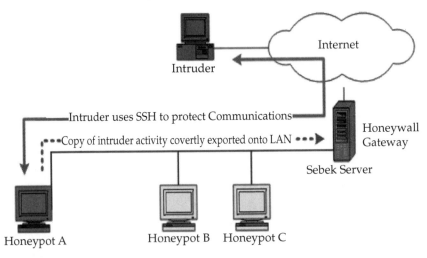

Figure 11.3 Sebek Architecture.

The client captures data off of a honeypot and exports it to the network which is collected by the server. The client resides entirely in kernel space on the honeypot and records all data via the read () system call. This data is then exported to the server over the net in a manner that is difficult to detect from the honeypot running Sebek. The server then gathers the data from all of the honeypots sending data.

Sebek now comes with a web based analysis interface. This interface provides users with the ability to monitor keystroke activity, search for specific activity, recover SCPed files and in general provides an improved data browsing capability.

For a skilled intruder who has intimate knowledge of the Operation System there are ways of detecting the presence of Sebek. Hence the future work will be to reduce the chances of detection. The other goal of Sebek is to expand the types of data collected and analyzed so that it can be used as a honeypot glass box analysis tool.

11.3 HONEYPOT BASED NETWORK FORENSICS FRAMEWORKS

Existing digital Forensics models involves various phases for carrying out network investigation. Figure 11.4 shows a Generic Framework for Network Forensics.

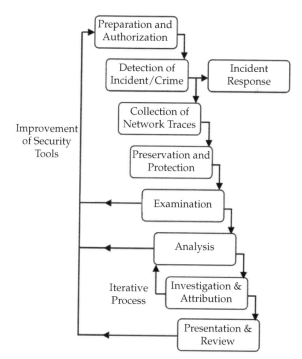

Figure 11.4 Generic Framework for Network Forensics.

11.3.1 Generic Process Model

The generic process model [18] for network forensic analysis aggregates many of the phases available in the existing digital forensic models but builds on those phases which are specific to networked environments only. Honeypots can be used in many of these phases for two major benefits—technical and ethnological [19, 20]. Honeypots can technically detect the presence of rootkits, Trojans and potential zero-day exploits. They can help ethnologically to gain understanding on the areas of interest and interaction between blackhat teams. Honeypot forensics focuses on understanding blackhat tools and techniques before and after the intrusion on the honeypot. Each of the nine phases in the general process model is explained below, while discussing the use of honeypots:

Phase1: Preparation and Authorization

Network forensics is applicable only to environments where network security tools (sensors) like intrusion detection systems, packet analyzers, firewalls, traffic flow measurement software are deployed at various strategic points on the network. The required authorizations to monitor the network traffic are obtained and a well defined security policy is in place so that privacy of individuals and the organization is not violated. There is a need to carefully setup the Honeypots to lure attackers and study their behavior. Honeypots are easy to fingerprint and a skilled blackhat may quickly realize the trap and immediately leave without leaving much evidence.

Phase 2: Detection of Incident/Crime

The alerts generated by various security tools, indicating a security breach or policy violation, are observed. All unauthorized events and anomalies noticed will be analyzed. The presence and nature of the attack is determined from various parameters. Snort can log network activity and all outgoing traffic from a honeypot will be suspicious. Quick validation is done to assess and confirm the suspected attack. This will facilitate the important decision whether to continue investigation or ignore the false alarm. The positive confirmation yields two directions—incident response and data collection.

Phase 3: Incident Response

The response to the crime or intrusion detected is initiated based on the information gathered to validate, assess and investigate the incident. The response initiated depends on the type of attack identified and is guided by organization policy, legal and nature of business. The response is initiated at two levels. First, an action plan on how to contain future attacks and recover from the existing damage in initiated. Second, a process to mitigate the attack and safe guard resources is launched, based on the conclusions arrived in the investigation phase (described below). The knowledge gained about the attack using the honeypots can help safeguard the unaffected live production systems. The steps followed by the intruder are recorded and become a signature to detect similar zero-day attacks in the future.

Phase 4: Collection of Network Traces

Data is acquired from the sensors used to collect traffic data. The sensors used must be secure, fault tolerant, have limited access and must be able to avoid compromise. Honeypots can be a useful tool to collect data in a stealth fashion. The integrity of data logged and network events recorded can also be ensured as the intruder is not aware that he is being monitored. The amount of data logged will be enormous requiring huge memory space and system must be able to handle different formats appropriately.

Phase 5: Protection and Preservation

The original data obtained in the form of traces and logs is stored on a back up device. A hash of all the trace data is taken and the data is protected. Chain of custody is strictly enforced so that there is no unauthorized use or tampering. Another copy of the data will be used for analysis and the original collected network traffic is preserved. The use of honeypots facilitates the presence of apparent data which does not have any commercial value but is sufficient to attract the intruder. However the information logged helps in learning the attack methodology but does not meet the legal requirements.

Phase 6: Examination

The traces obtained from various security sensors are integrated and fused to form one large dataset on which analysis can be performed. Mapping and time lining of this data is also performed. This is done so that crucial information is not lost or mixed up. Data hidden or camouflaged by the attacker needs to be recovered. Encrypted data poses a major challenge and honeypot tools like Sebek capture the data after the attacker decrypts it on the victim machine. Traffic originating out of the Honeypots need special focus as they don't provide any real services. Redundant information and unrelated data is removed and meaningful events are pinpointed. Various network flows involving long period of time and exchange of large number of bytes can be identified.

Phase 7: Analysis

The evidence collected is searched methodically to extract specific indicators of the crime. The indicators are classified and correlated to deduce important observations using the existing attack patterns. Statistical and data mining approaches are used to search the data and match attack patterns. The attack patterns are put together and the attack is reconstructed and replayed to understand the intention and methodology of the attacker. Honeypots can be used to build timeline of events from the network trace obtained by an IDS like Snort and correlate the same with a timeline built from system forensics using a tool like Sebek. Statistics on protocols in the captured data are also instructive about the various compromises attempted.

Phase 8: Investigation and Attribution

The information obtained from the evidence traces is used to identify who, what, where, when, how and why of the incident. This will help in source traceback, reconstruction of the attack scenario and attribution to a source. The most difficult part of the network forensic analysis is establishing the identity of the attacker. Two simple strategies of the attacker to hide himself are IP spoofing and stepping stone attack. Researchers have proposed many IP traceback schemes to address the first attack (mostly applicable only for DDOS attacks) and is still an open problem. Stepping stones are created by attackers to use compromised systems to launch their attacks. They can be detected using similarity and anomaly based approaches applied to packet statistics. Honeypots can be used in attacker profiling and to understand the intentions. The evidence collected by the honeypot is not legally valid as the honeypot is not a real system.

Phase 9: Presentation and Review

The observations are presented in an understandable language to the organizations management and legal personnel while providing explanation of the various standard procedures used to arrive at the conclusion. The systematic documentation is also included to meet the requirements. The statistical data is interpreted in support of the conclusions arrived. A thorough review of the incident is done and counter measures are recommended to prevent similar incidents in future. The results are documented to influence future investigations and in improvement of security products.

11.3.2 Honeypot Based Frameworks for Forensics

Honeypot frameworks are used to attract the attackers so that their process methodology can be observed and analyzed to improve defense mechanisms. Some of these implementations are:

1. Honeytraps

To collect information about blackhat activities and to learn about their techniques, Yasinsac and Manzano developed the Honeytraps [21] framework. It is a deception tool with protection and defense mechanisms formulated. They are Honeypot or Honeynet systems which attract intruders to enter the host by emulating a known vulnerability. Once an attacker penetrates a honeytrap, data is captured to detect and record his actions. This data can be used to profile the tools and tactics used by the attackers putting the investigators into an offensive mode.

Two architectures, serial and parallel, as shown in Figure 11.5 and Figure 11.6, facilitate the forensic investigation. The serial architecture places the honeytrap between the Internet and the production system. Recognized users are filtered to the production systems and the blackhats are contained in the honeytrap. The parallel architecture allows the honeytrap to be independent of the production system. Once the system detects the presence of blackhat, the forensic alert system is activated.

If the attack is detected, forensic processes are activated on the honeytrap and production systems. Once the attack is contained, the investigation process is begun to determine the identity of the intruder on the production system.

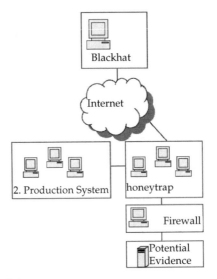

Figure 11.5 Serial Architecture.

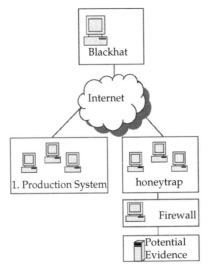

Figure 11.6 Parallel Architecture.

2. Honeynet based Distributed Adaptive Network Forensics and Active Real Time Investigation

A *Distributed Adaptive Network Forensics* Framework was developed by Ren and Jin [22], that was based on Honeynet and did the active Real Time Investigation. Honeynet system is used to lure the attackers and to gain information about new types of intrusions. Network forensics system then, analyzes and reconstructs the attack behaviors [23]. These two systems when integrated can help build an active, self-learning response system to profile the intrusion behavior and investigate the attack original source.

There are four elements in the network forensics system—server, agents, monitor, and investigator. Agents are engines of the data gathering, extraction and secure transportation. They are deployed on the monitored host and network. Server integrates the forensics data from agents and analyzes it. Monitor is a packet capture machine to adaptively capture the network traffic. Investigator surveys the network provides the mapping topology. The server guides the network packet filter and captures behavior. It launches the investigation program in response to the attacks. It can launch the real time investigation also in response to the network intrusion.

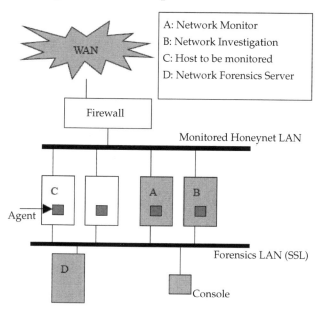

Figure 11.7 Honeynet based Distributed Adaptive Network Forensics System.

3. A framework for attack patterns' discovery in Honeynet data

Another framework was developed for attack patterns' discovery in Honeynet data ped by Thonnard and Dacier [24]. The aim of this framework is to find groups of network traces sharing various kinds of highly similar patterns within an attack data set. The design has a flexible clustering tool to analyze the time series of the attacks. Malicious network traffic is obtained from the distributed set of Honeynet responders. Time signature is used as a primary clustering feature and attack patterns are discovered using attack trace similarity. Attacks are detected as a series of connections, zero-day and polymorphic attacks are detected based on similarity to other attacks and knowledge from the Honeynet data is used in intrusion detection efforts. The clustering method does feature selection and extraction, defines a pattern proximity measure and group similar patterns. The result of the clustering applied to time series analysis enables detection of worms and botnets in the Honeypot traffic.

4. Automated Network Forensic Tool

An automated analysis framework was designed by Merkle [25] to investigate the of network based evidence in response to cyberspace attacks. The two major challenges of network forensics, namely the *complexity* problem of analyzing raw traffic data and the *quantity* problem of amount of data to analyze are addressed in this framework. The model integrates results of data logged by various tools into a single system that exploits computational intelligence to reduce human intervention. This integrated tool is referred as the automated network forensic tool.

An isolated network of virtual machines is built into a Honeynet. A hybrid evolutionary algorithm is implemented on the attacker system so that it generates reasonably realistic variations on known network attacks against the Honeynet. The resulting network forensic data is collected to train the automated network forensic tool. A set of open source forensic tools like tcpdstat (statistical data), Snort (IDS alerts) and Argus (statistical analysis of session data) are used to build the integrated system.

The network forensic analysis consists of many phases as seen in the generic process model. Various tools are used in each stage and the information produced in one stage is characterized and transformed for use by other tools in the succeeding stages. Time consuming and error prone processes are identified and automated applying computational intelligence techniques. The Honeynet data sets are randomly partitioned to train the system. Each component and the integrated system is tested against the datasets to evaluate the efficiency of the tool.

In this way, many Honeypot based frameworks have been designed for Network Forensics. Honeypot and Forensics work together to attracts the attackers and then to analyze their process methodology to improve the defense mechanisms.

Limitations

There are many limitations associated with different phases of the process model and framework implementations. Some of these are:

Collection and Detection

The first step in network forensic analysis involves collection of network traffic and detection of attacks from the logs. The challenge is to identify useful network events and record minimum representative attributes for each event so that the least amount of information with the highest probable evidence is stored [26]. This results in reduction of data storage requirements. A data digest will be sufficient for discovery of malicious behavior and a full capture is required for reconstruction of attack behavior.

Data Fusion and Examination

The data captured from various tools must be aggregated and examined to ascertain whether investigation should be commenced. Data fusion of all the logs collected from various security tools deployed in each hosts on the entire network is a crucial problem [27]. The dependencies of packet attributes from various tools and reconnaissance of the attributes from different hosts validate an attack. Characterization of anomalous network events and distinguishing attack traffic from legitimate traffic by searching for patterns of anomalies is the major challenge.

Analysis

The critical step in the entire process of network forensics is to analyze the attack data and arrive at a conclusion, pointing at the source. Classification and clustering of network events is done, so that scrutiny of large volumes of data to understand their relationship with attacks becomes easy. Parsing and analysis of complex protocols also needs focus. Pattern recognition of anomalies using soft computing and data mining techniques can be applied for classification, correlation and link analysis. The categorization of attack patterns and attack reconstruction in order to understand the intention and methodology of the attacker needs research focus [28].

Investigation

The investigation must enable the attribution of an attack to a host or a network. The results must meet the admissibility criteria in a court of law. The analysis of logs and other network traces must lead to the source of attacks. IP Traceback involves tracing back to the source address of the attacker overcoming IP spoofing [29]. Detecting and profiling TCP connection chains can bring out the stepping stones used to launch an attack. Creating a topology database and IP location mapping to locate an attacker geographically is a major challenge.

Incident Response

Active real time response to the network misuse is to be performed so that important data is not lost by the time response is initiated. The response processes are to be launched immediately when the alerts begin. The key issue to be maintained is that the attacker must not be aware of the response [30].

11.4 SUMMARY

Network forensics is the investigation of the attacks by tracing the attack back to the source and attributing the crime to a person, host or a network. It has the ability to predict future attacks by constructing attack patterns from the existing traces of intrusion data. The incident response to an attack is much faster. The preparation of authentic evidence, admissible in the legal system, is also facilitated. This Chapter makes an extensive survey on various network forensic framework implementations which are based on honeypots. The functionality of a few NFATs and NSM tools is also discussed while highlighting Sebek, a honeypot based tool. A generic model for network forensics is described, discussing the use of honeypots in its various phases. Some of the limitations and challenges in network forensics as an alternate approach to security are also presented. There is a need of urgently addressing these limitations and challenges in various tools and framework implementations so that the perpetuators of cyber crime are traced back and prosecuted. This will act as a deterrent, resulting in drastic decrease in network crime rate, thereby improving security.

EXERCISES

Short Answer Questions:

1. Explain the concept of Network Forensics.
2. What are the various classifications of Network Forensics Systems?
3. What is the need of Honeypot in Network Security?
4. Explain the role of Honeypot in Netowork Forensics.
5. Explain SEBEK as Network Security Tool.

Long Answer Questions:

1. What are the various Network Forensics Analysis Tools?
2. How Honeypot is used in various phases of Generic Network Forensics Model?
3. Explain some of the Honeypot based Network Forensics Framework.
4. What are the limitations of Network Forensics Process Model? Give some solution to overcome them.

REFERENCES

[1] V. Broucek and P. Turner, "Forensic computing: Developing a conceptual approach for an emerging academic discipline," in *5th Australian Security Research Symposium*, 2001.

[2] H. Berghel, "The discipline of Internet forensics," *Communications of the ACM*, vol. 46, p. 20, 2003.

[3] G. Palmer, "A road map for digital forensic research," in *First Digital Forensic Research Workshop*, Utica, New York, 2001, pp. 27–30.

[4] M. Ranum, "network forensics " in http://www.ranum.com.

[5] W. Ren and H. Jin, "Modeling the network forensics behaviors," in *Workshop of the 1st International Conference on Security and Privacy for Emerging Areas in Communication Networks*, 2005, pp. 1–8.

[6] S. Garfinkel, "Network Forensics: Tapping the Internet," *IEEE Internet Computing*, vol. 6, pp. 60–66, 2002.

[7] S. Perry, "Network forensics and the inside job," *Network Security*, vol. 2006, pp. 11–13, 2006.

[8] "ISO/IEC 27001: 2005," in http://www.iso.org/iso/catalogue_detail.htm?csnumber=42103.

[9] "netForensics Security Compliance Management," in http://www.netforensics.com/compliance.

[10] "Honeynet Project: Know Your Enemy: Honeynets—What a honeynet is, its value, how it works, and risk involved," in http://old.honeynet.org/papers/honeynet/.

[11] L. Spitzner, "Know Your Enemy: Defining Virtual Honeynets," in http://www.honeynet.org/.

[12] D. Moore, C. Shannon, G.M. Voelker, and S. Savage, "Network telescopes: Technical report," *CAIDA, April,* 2004.

[13] W. Harrop and G. Armitage, "Defining and evaluating greynets (sparse darknets)," in *Local Computer Networks 2005. 30th Anniversary. The IEEE Conference on*, Sydney,NSW, 2005, pp. 344–350.

[14] R. Sira, "Network Forensics Analysis Tools: An Overview of an Emerging Technology," *GSEC, Version,* vol. 1, 2003.

[15] V. Corey, C. Peterman, S. Shearin, M.S. Greenberg, and J. Van Bokkelen, "Network forensics analysis," *IEEE Internet Computing*, vol. 6, pp. 60–66, 2002.

[16] E.S. Pilli, R.C. Joshi, and R. Niyogi, "Network forensic frameworks: Survey and research challenges," *Digital Investigation*, vol. 7, pp. 1–12, April 2010.

[17] L. Spitzner, "Know Your Enemy: Sebek—A kernel based data capture tool," in http://www.honeynet.org/papers/sebek.

[18] E.S. Pilli, R.C. Joshi, and R. Niyogi, "A Generic Framework for Network Forensics," *International Journal of Futuristic Computer Applications*, vol. 1, 2010.

[19] F. Raynal, Y. Berthier, P. Biondi, and D. Kaminsky, "Honeypot forensics part I: Analyzing the network," *IEEE Security and Privacy,* vol. 2, pp. 72–78, 2004.

[20] F. Raynal, Y. Berthier, P. Biondi, and D. Kaminsky, "Honeypot forensics, part II: analyzing the compromised host," *IEEE Security and Privacy,* vol. 2, pp. 77–80, 2004.

[21] A. Yasinsac and Y. Manzano, "Honeytraps, a network forensic tool," in *Proceedings of 6th Multi-Conference on Systemics, Cybernetics and Informatics* Florida, USA, 2002.

[22] W. Ren and H. Jin, "Honeynet based distributed adaptive network forensics and active real time investigation," in *Proceedings of the 2005 ACM symposium on Applied computing,* Saanta Fe, New Mexico, 2005, pp. 302–303.

[23] S. Riebach, E.P. Rathgeb, and B. Toedtmann, "Efficient deployment of honeynets for statistical and forensic analysis of attacks from the internet," *IFIP NETWORKING 2005,* vol. 3462, pp. 756–767, 2005.

[24] O. Thonnard and M. Dacier, "A framework for attack patterns' discovery in honeynet data," *Digital Investigation,* vol. 5, pp. 128–139, 2008.

[25] L.D. Merkle, "Automated network forensics," in *Proceedings of the 2008 Genetic And Evolutionary Computation conference (GECCO,2008),* Atlanta, GA, USA, 2008, pp. 1929–1932.

[26] S. Mukkamala and A.H. Sung, "Identifying significant features for network forensic analysis using artificial intelligent techniques," *International Journal of Digital Evidence,* vol. 1, pp. 1–17, 2003.

[27] W. Ren, "A Network Forensics Model for Information Security," in *Proceedings of 3rd International Conference on Information System Technology and its Applications (ISTA 2004),* Utah, USA, 2004, pp. 229–234.

[28] A. Almulhem, "Network Forensics: Notions and Challenges," in *Proceedings of 9th IEEE International Symposium on Signal Processing and Information Technology (ISSPIT 2009)* UAE, Dec 2009.

[29] S. Mitropoulos, D. Patsos, and C. Douligeris, "Network Forensics: Towards a classification of traceback mechanisms," in *Proceedings of 1st Inernational Conference Security and Privacy for Emerging Areas in Communication Networks,* 2005, pp. 9–16.

[30] H. Khurana, J. Basney, M. Bakht, M. Freemon, and R. Butler, "Palantir: a framework for collaborative incident response and investigation," in *Proceedings of 8th Symposium on Identity and Trust on the Internet,* MaryLand, 2009, pp. 38–51.

Index